An ENVIRONMENTAL HISTORY of the CIVIL WAR

CIVIL WAR AMERICA

*Peter S. Carmichael, Caroline E. Janney,
and Aaron Sheehan-Dean, editors*

This landmark series interprets broadly the history and culture of the Civil War era through the long nineteenth century and beyond. Drawing on diverse approaches and methods, the series publishes historical works that explore all aspects of the war, biographies of leading commanders, and tactical and campaign studies, along with select editions of primary sources. Together, these books shed new light on an era that remains central to our understanding of American and world history.

An ENVIRONMENTAL HISTORY of the CIVIL WAR

JUDKIN BROWNING & TIMOTHY SILVER

THE UNIVERSITY OF NORTH CAROLINA PRESS

Chapel Hill

© 2020 The University of North Carolina Press

All rights reserved

Designed by Jamison Cockerham
Set in Arno, Scala Sans, Rudyard, Ashwood, Brothers, and Dear Sarah
by Tseng Information Systems, Inc.

Cover illustrations: Map of the siege of Vicksburg, 1864, by Adam Badeau, 1885; and *Stuck in the Mud: A Flank March across Country during a Thunder Shower*, by Edwin Forbes, ca. 1876; both courtesy of the Library of Congress.

Manufactured in the United States of America

The University of North Carolina Press has been a member of the Green Press Initiative since 2003.

LIBRARY OF CONGRESS CATALOGING-IN-PUBLICATION DATA
Names: Browning, Judkin, author. | Silver, Timothy, 1955– author.
Title: An Environmental History of the Civil War / Judkin Browning and Timothy Silver.
Other titles: Civil War America (Series)
Description: Chapel Hill : The University of North Carolina Press, [2020] | Series: Civil War America | Includes bibliographical references and index.
Identifiers: LCCN 2019041157 | ISBN 9781469655383 (cloth) | ISBN 9781469655390 (ebook)
Subjects: LCSH: Nature—Effect of human beings on—United States—History—19th century. | United States—History—Civil War, 1861–1865. | United States—Environmental conditions—History—19th century.
Classification: LCC E468.9 .B883 2020 | DDC 973.7—dc23
LC record available at https://lccn.loc.gov/2019041157

CONTENTS

Introduction: More than the Mud March   1

one **SICKNESS** 9
Spring–Winter 1861

two **WEATHER** 39
Winter 1861–Fall 1862

three **FOOD** 71
Fall 1862–Summer 1863

four **ANIMALS** 102
Summer 1863–Spring 1864

five **DEATH AND DISABILITY** 133
Spring 1864–Fall 1864

six **TERRAIN** 160
Fall 1864–Spring 1865

Epilogue: An Environmental Legacy   187

Acknowledgments   201

Notes   205

Bibliography   227

Index   251

ILLUSTRATIONS

The Mud March 3

Camp life 20

Hospital ward 22

Quinine in whiskey 30

Butler and Lincoln cartoon 33

California flood 40

Map of Sibley campaign 45

Attack on Fort Henry 48

Map of Virginia Peninsula 55

March from Williamsburg 57

Bridge over the Chickahominy River 60

Foraging in Virginia 73

Cornfield at Antietam 78

Confederate conscription 84

Food in Ohio 90

Map of Vicksburg and vicinity 95

Dead horses at Gettysburg 104

Giesboro Depot 108

Burning dead horses at Fair Oaks   *120*

"Reinforcements for Our Volunteers"   *123*

Beef for the Army — on the March   *129*

Battle of Spotsylvania   *135*

Wounded at the Wilderness   *141*

Confederate dead at Spotsylvania   *144*

"Air-Tight Deodorizing Burial-Case"   *146*

Embalming   *148*

Field hospital   *154*

Andersonville prison   *157*

Saltworks in Saltville, Virginia   *163*

Wilderness battlefield   *171*

Forts at Atlanta   *176*

Winter camp in Virginia   *178*

Mirror Lake in Yosemite Park   *198*

An ENVIRONMENTAL HISTORY of the CIVIL WAR

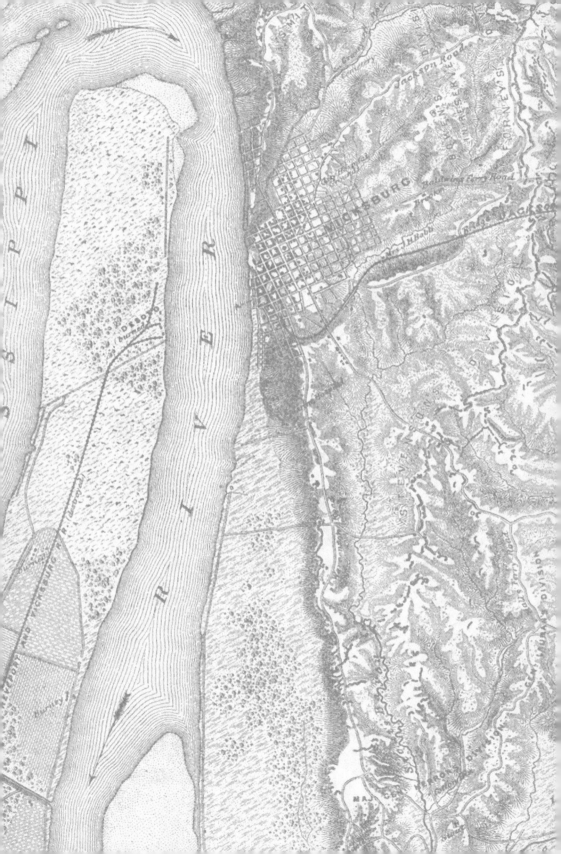

INTRODUCTION

MORE THAN THE MUD MARCH

The day after Christmas 1862, Union general Ambrose Burnside started planning a bold maneuver to capture the Confederate capital of Richmond. The commander called for his Army of the Potomac to cross Virginia's Rappahannock River on pontoon bridges west of Fredericksburg—a city halfway between Washington, D.C., and Richmond—and move south around the Confederate troops on the other side. From that position, Burnside hoped to squeeze the enemy between his army and the banks of the Rappahannock. With southern forces defeated or on the defensive, Burnside believed he might have an easy path to Richmond. Once there, he would deal the South a blow from which it could not recover. Such an aggressive move might also boost northern morale, which had been at low ebb since a humiliating defeat at Fredericksburg two weeks earlier. Few subordinate officers expressed much faith in Burnside or his plan. At least two generals met with President Abraham Lincoln to say as much, and for a moment, the commander in chief wavered. But when Burnside threatened to resign, Lincoln reluctantly approved the action.[1]

During the first weeks of 1863, Burnside watched the weather. Sunny skies and cool temperatures allowed local roads to dry out and become firm enough for travel. Even a quick cold snap on January 17 seemed to have little effect on his men or his projected path to the Rappahannock. Encouraged by the good weather, the general put his army in motion at noon on January 20. As he approached the river, however, Burnside got word of Rebel troops stationed south of the river in a position he had not anticipated. He decided to delay the crossing for a day while he considered the latest intelligence. After that, he intended to move his two corps—Burnside preferred to call them "Grand Divisions"—consisting of 75,000 men into position for an attack on Robert E. Lee's Confederate forces.[2]

Around 9:00 on the night of January 20, just as the first of his men reached the river's edge, a "cold, heavy, incessant" rain set in along the Rappahannock. At times the rain mixed with snow, and the precipitation did not stop for thirty hours. Driving wind blew down hastily pitched tents. The relentless downpour doused campfires, leaving shivering soldiers no protection against the elements. Gen. Régis de Trobriand recalled that nightfall brought "a funereal aspect, in which the enthusiasm is extinguished." In short order, the once-firm Virginia roads became a muddy morass of ruts and ditches as impassable as any swamp. Trobriand asserted, "the mud is not simply on the surface, but penetrates the ground to a great depth." According to one of Burnside's artillery officers, "The mud was so deep that sixteen horses could not pull one gun." Wagons stuck up to their hubs, draft animals died from exhaustion, and mules "drowned in the middle of the road." At least four of the pontoon bridges needed to cross the river remained mired firmly in the muck. Theodore A. Dodge, a soldier in the 110th New York who followed Burnside to the river, explained it this way: "Mud is really king. He sets down his foot and says, 'Ye shall not pass,' and lo and behold we cannot."[3]

Something about the Virginia quagmire proved especially insidious. Dodge could not put his finger on it, but for some reason, he believed, "mud wields a more despotic sway these last two days than I ever saw him wield before." Unable to go any farther, Burnside's bedraggled men forlornly returned to their winter camps near Fredericksburg without so much as firing a shot. Meanwhile, on the other side of the river, Lee's men looked on with glee, posting large signs that read "This Way to Richmond" (complete with arrows pointing in the opposite direction from the capital city) and "Burnside Stuck in the Mud."[4]

In the aftermath of the fiasco, Burnside blamed "insubordinate" Union officers for his woes and threatened to resign unless Lincoln dismissed the alleged offenders. The officers and their supporters, some of whom coveted Burnside's command, pointed the finger at the general. Eventually Lincoln, never enamored with Burnside's effort in the first place, convinced the general to accept another assignment and entrusted the Army of the Potomac to Joseph Hooker, one of Burnside's critics. Perhaps better than any other incident, Burnside's ill-advised "Mud March" demonstrates the indecision, lack of leadership, and internal rancor that plagued the Union high command in late 1862 and early 1863.[5]

For those who write the military history of the Civil War, the Mud March is one place (and sometimes the only place) where the natural world — in the

Sketch of the Union Army of the Potomac during the "Mud March," January 21, 1863. Courtesy Library of Congress.

form of rain and bottomless Virginia mud—becomes important to the narrative. Tellingly, it is the only campaign named for a weather-induced feature, rather than a geographic one. The horrendous weather that overtook Burnside has prompted a few scholars to look into the atmospheric conditions that brought on the storm and the soils that so quickly became a quagmire. But most military historians come back to Burnside, citing his overly ambitious plan, his failure to understand the folly of moving troops in winter, and the one-day delay that kept him from crossing the river. Unlike Theodore Dodge and other soldiers who lived through those difficult days in 1863, military historians have been much quicker to note the foibles of General Burnside than to acknowledge the tyranny of King Mud.[6]

Over the last two decades, environmental historians, a group of scholars generally more attuned to the role of nature in human endeavors, have called for a different approach. As Jack Temple Kirby explains it, people are "connected creatures, obligated partners in a dynamic natural community." Within that community, the natural world always affects any human activity. In turn, human actions, in war or any other enterprise, alter nature. When it comes to the Civil War, some environmental historians have viewed the conflict as a struggle for resources; others have called attention to the health of soldiers, civilians, slaves, and freedpeople. A few have focused on the wanton destruction of the natural and built environments and the ways Americans reacted to the unprecedented devastation. The impact of the war on agriculture, especially in the South as it reorganized after emancipation, has also drawn scrutiny. Lisa M. Brady, one of the first environmental historians to examine the Civil War, has investigated the ways military strategy reflected prevailing ideas about nature and how those ideas influenced individual campaigns.[7]

Even so, we still lack a work that considers the four years of war—the musters, training, troop movements, battles, home front, and aftermath—in an environmental context. This book, a collaborative effort between a military and an environmental historian, is an attempt to write such a history. Simply stated, we have tried to reimagine the war, not just as a military action but also as a biotic or biological event, one crucial to the history of the American environment. Decades ago, Alfred W. Crosby, a pioneer in environmental history, reminded us that human beings are never really alone in the natural world. They live side by side with what Crosby, in a marvelous turn of phrase, called the "portmanteau biota": the conglomeration of microbes, crops, weeds, and domestic animals that reside in their bodies and on their farms and fields. When Europeans settled on other continents, Crosby argued, they and their associated organisms were not only colonizers but also agents of biological upheaval that transformed both people and nature. This mass relocation of people, plants, and animals took place in a natural world that was not static but one in which climate, weather, winds, and a host of other factors helped determine the pace and extent of change.[8]

Similarly, we treat the war as an ecological event that not only affected people but also altered natural systems and reshaped the already complex interaction between humans, other organisms, and the physical environment. Such a history requires merging traditional military sources with material from relevant sciences, scholarly territory often unfamiliar to historians. This approach also grants agency to the natural world, not as the sole determinant of events but as a prominent and often neglected actor in a complicated story. As environmental historian Ellen Stroud writes, paying attention to the "material stuff of nature"—rain, dirt, bacteria and viruses, animals, and human bodies—does not mean that one ignores human action and decision-making. Instead, giving equal time to nature provides a new context, a means "of telling better histories," a way "to bring to light connections, transformations, and expressions of power that otherwise remain obscured."[9]

Viewed in this context, the "Mud March" becomes much more than a confrontation between Burnside, his officers, rain, and soil. Before he could move on Richmond, the general had to make sure he had adequate provisions to feed his men and that they were healthy enough to embark on a campaign. Likewise, every horse and mule that pulled a wagon or moved artillery toward the Rappahannock had to receive fodder and care. All those plans went awry with the bad weather. The prolonged exposure to the elements and the physical exertion it required weakened the soldiers. Carcasses from the animals that suffocated in the mud had to be disposed of or left to rot, adding to the misery of the rain-soaked camps. Additionally, every man who marched with Burnside already carried within his body various microorganisms that could cause sickness in the right conditions. Those conditions flourished during this cold, wet slog. As a result, disease ran rampant through the dispirited Union ranks, exacerbated by overcrowding, poor nutrition, and poor sanitation. When General Hooker finally took over, he commanded a demoralized and diseased army wrecked by poor human decisions in difficult environmental circumstances. "I do not believe I have ever seen greater misery from sickness than now exists in the Army of the Potomac," wrote the army's medical inspector general. One medical officer suggested that the "Mud March" cost as many lives as the Battle of Fredericksburg by the time the illnesses had run their courses. For the rest of their lives, numerous soldiers believed that their chronic joint pain and bowel complaints stemmed directly from those three frigid, wet, and muddy days along the Rappahannock. Disheartened by their experience, more men deserted the Federal ranks that bleak winter than at any other time of the war.[10]

What is true of the Mud March is true of the war as a whole. Soldiers from rural areas crowded together in training camps, creating a new and inviting environment for the microorganisms living in their bodies. Armies larger than many American cities confronted each other on the confines of the battlefield, bringing to rural areas all the problems of sanitation and waste disposal associated with urban life. Thousands of animals accompanied the troops—horses and mules that moved men, artillery, and supplies, as well as cattle and hogs that provided meat for sustenance. Along with people, the animals were part of a massive mammalian migration that had enormous implications for the natural world. Dead animals and dead people had to be disposed of, sometimes on a massive scale. Peculiarities of terrain often dictated what armies could do, and the armies, in turn, altered the land they occupied. As historian Stephen Berry writes, "The Civil War was a massive stir of the biotic soup, and in many ways *that stir*, more than the battles themselves, was the real story of the war." It also constituted, we would add, a significant episode in the changing story of the American environment.[11]

Rethinking the Civil War in these terms immediately presents a problem of chronology. For the humans who fought in it, the Civil War began in April 1861 and ended in April 1865. Key battles and campaigns are easily identifiable. The natural world, however, moves to its own rhythms, influenced but not bound by human notions of time and space. Recognizing that distinction but firmly believing that modern readers prefer linear stories with a beginning and an end, we have made the difficult choice to focus on certain environmental themes during specific seasons of war.

In chapter 1, we examine the health of soldiers in the first six months of the conflict. We explore how the assembly of thousands of troops in preparation for war caused outbreaks of infectious disease, and we delve into the multiple disease environments created by the various campaigns. We also discuss other factors that affected the bodies of the volunteers, such as marching in the summer heat, as well as the quality and quantity of the food and water that soldiers received. In chapter 2 we focus on how the weather of 1862 shaped military campaigns from California to Virginia. Floods emerged in the midst of a decade-long drought, forcing commanders, soldiers, and civilians to make decisions that dramatically affected the direction of the war. We situate chapter 3 in the year from the summer of 1862 to the summer of 1863 to analyze how both sides provided food for their armies and civilians amid the increasingly destructive conflict. The lack of food influenced the origins and conclusions of campaigns from Antietam to Vicksburg.

Animals take center stage in chapter 4, as we follow the plight of horses, cattle, and hogs from the summer of 1863 through the winter of 1864. These animals provided the engines and protein for the armies, and the war brought suffering, disease, and wholesale death to them just as it did to humans. In chapter 5, we examine the death and disability of soldiers during the spring and summer of 1864 — grim seasons dominated by the brutal Overland Campaign in Virginia. The armies struggled for supremacy while confronting myriad problems created by tens of thousands of wounded and dying men. In chapter 6, we look at the various landscapes on which battles took place from the fall of 1864 until the spring of 1865. We examine how terrain influenced the fighting and how battles altered the land, including something as basic as the quest for salt, to monumental military conquests like the capture and destruction of Atlanta. In the epilogue, we discuss the environmental legacy of the war and the ways it continued to shape Americans' relationship with nature after Appomattox.

No organization scheme is perfect. Infectious diseases endured well beyond 1861; turbulent weather did not suddenly stop in 1862; people and animals died long before 1864. With that in mind, we necessarily stretch our chronological parameters, looking forward or backward a bit in each chapter to provide context and complete the story. With more than 50,000 books currently available on the Civil War, we are also keenly aware that no important battle, leader, or tactic has escaped scrutiny from scholars. We make no claim to undiscovered sources or unrecorded events. Instead, what we offer is a more holistic way of thinking about the Civil War, one that did not escape some of the most astute observers of people and nature in the nineteenth century. When Herman Melville sat down to compose a poem about the bloody battle at Malvern Hill along Virginia's James River in July 1862, he turned his attention not to strategy or tactics but the natural world. He took note of surrounding forests where "rigid comrades" lay in death and spoke of the "leaf-walled ways" that allowed for the passage of men, animals, and weapons of war. But Melville also seemed to recognize that, whatever the plans of troops and generals, nature stubbornly went its own way, often oblivious to human designs. "The elms of Malvern Hill," he wrote, "Remember everything / But sap the twig will fill / Wag the world how it will / Leaves must be green in spring." So it is in war, as nature and people influence each other. And so it will be as we examine the events of the Civil War and its environmental legacy, starting with the illnesses that emerged as thousands of young men flocked to join the great adventure in 1861.[12]

One
SICKNESS
SPRING–WINTER 1861

On April 24, 1861, eleven days after Fort Sumter surrendered, eager Confederate recruits gathered in Rocky Mount, North Carolina, a sleepy village on the Tar River in the eastern coastal plain. Secession sentiment ran strong in the Edgecombe County town. Since January, residents had attended rallies, listened to speeches, and raised a lone star flag in support of the Confederate cause. On that April day, seized by *rage militaire*, 83 local young men volunteered to serve in a company of state troops—four weeks before North Carolina formally seceded from the Union. One month later a second company formed with 93 enthusiastic enlistees. Like many new soldiers, some of those troops probably worried that the war might end before they could take part in the great adventure. These 176 recruits rejoiced when orders finally came to depart for training. A few weeks later, the men arrived at a military camp in Garysburg, North Carolina, near the Virginia border. There they joined another thousand or so recruits from eight other Carolina counties. Together they formed the Fifth North Carolina Volunteers and began to learn the basic skills of war. Similar scenarios played out in hundreds of towns and cities across the North and South that spring as local volunteer companies became building blocks of the Union and Confederate armies. Everywhere, it seemed, men were on the move, preparing for an endeavor that would change their lives.[1]

The initial musters also brought immediate and profound changes in the natural world. The large gatherings of humans provided a new and nearly ideal environment for an invisible organism that quickly took up residence in the noses and throats of many recruits. Within days, the microbe made its presence felt among the Rocky Mount contingent. At first, a few men reported feeling tired and sluggish. Within a day or so, high fever left some of them shivering, even in the warmth of a Carolina spring. A bright red skin rash followed, appearing first at the hairline and spreading over the rest of the body. At that point, medical personnel realized that the rash likely signaled the presence of measles, but doctors could do little to protect the new soldiers from what soon became a raging epidemic. One soldier remarked, "I had thought as a matter of course, every grown man had had the measles when a boy, but in this I have made a grave mistake." Known to physicians as *Rubeola*, measles results from a virus that lives in human mucus. The microbe needs people to survive and propagate but, because infection also confers a measure of immunity, the virus cannot sustain itself for long without a steady supply of fresh nonimmune humans.[2]

By 1860, measles had become a fact of life in many American cities. Residents frequently contracted the disease in childhood when the body's natural defenses were better equipped to fight off the virus and provide immunity. City populations also tended to be mobile and transient, supplying *Rubeola* with fresh adult victims. The virus had more trouble establishing itself in rural America, where smaller and more stable human populations provided fewer children and nonimmune adults. Over time, epidemics became more infrequent until the virus eventually disappeared from small communities. Country people were often healthier than their urban counterparts, but freedom from rash and fever came at a biological price, as rural folk also lacked the immunity that some city folk enjoyed.[3]

When both sides mustered troops, small communities like Garysburg, North Carolina, became instant cities, populated by thousands of men in training for war. On the Confederate side, many of those who rushed to join up came from rural areas where they had not been previously exposed to measles. It took only one infected individual to release the virus into a large population of nonimmune hosts. Once set loose in that environment, measles became especially virulent, resulting in a longer than normal recovery time for those who survived. Doctors on both sides knew that the disease spread through contact and that quarantining victims might save others from infection. Even so, it could be difficult to identify measles in its early stages. Victims first suffered from a cough, runny nose, or conjunctivitis— symptoms common to a variety of ailments. Though such persons were contagious, the initial absence of the telltale rash, which usually did not appear until several days later, made it difficult to determine which soldiers should be quarantined and for how long.[4]

As preparations for battle transformed the relationship between virus and host, measles quickly became a scourge of Confederate training camps.

The disease broke out among soldiers at Richmond and Raleigh. At Camp Moore, near Kentwood, Louisiana, an entire garrison fell victim to the malady. At another camp, 4,000 of 10,000 new soldiers came down with measles at the same time. As one Virginia recruit recalled years later, "It seemed that half or more of the army had the [measles] the first year of the war." By autumn, some 8,000 would-be Confederate soldiers had been infected, including one out of every seven in the Army of Northern Virginia.[5]

Things were only slightly better among Union recruits. Wherever groups of men assembled, measles could be a threat. Serving with the Army of the Potomac, chief surgeon Charles S. Tripler noted, "In many of our regiments [measles] broke out before [soldiers] left their homes.... Some were more severely scourged than others, but nearly all suffered to some extent." The disease ran rampant through Union camps in the Midwest as well. From Cairo, Illinois, in November, Gen. Ulysses S. Grant wrote that more than 2,200 of his 18,000 soldiers were sick with measles, while Gen. Henry Halleck wrote from St. Louis in December that among his new recruits, "measles are prevailing and daily increase our sick list."[6]

Officers and medical personnel knew that if a soldier could endure the symptoms and survive the disease, he gained a measure of immunity. Treatment usually consisted of keeping victims as comfortable as possible and waiting for the sickness to run its course. Surgeons on both sides also encouraged the use of "seasoning camps," where new arrivals could be isolated from the general population and prepared for the battlefield. Such preventive measures helped, but given the peculiarities of *Rubeola* and the delayed onset of identifiable symptoms, some soldiers fell sick even in quarantine.[7]

Measles can kill the nonimmune, but more commonly fatalities result from pneumonia or another secondary infection that sets in as a weakened body fends off the virus. Overall, about one of every fifteen soldiers infected probably died from the contagion and its complications. As historian G. Terry Sharrer has suggested, more Confederate soldiers perished from measles than fell to northern gunfire in 1861. That same year, Union surgeons reported 21,676 cases of *Rubeola* and attributed 551 deaths to the contagion. Though Union figures may not account for all who died of various complications, the lower death rate suggests that measles took a greater toll among the predominantly rural southern soldiers.[8]

Putting people in motion in preparation for war unleashed other invisible shockwaves in the microbial world. During the first fiscal year of war, the Army of the Potomac reported 1,786 cases of mumps. Bacterial infections such as whooping cough (sometimes called "chin cough") and diphtheria (often lumped with strep and other ailments as "contagious diseases of the throat") surfaced in Union camps, along with other viral maladies such as chicken pox. Though specific evidence is more difficult to ascertain, southern soldiers probably suffered from all of those and more. According to Confederate officer John B. Gordon, "The large number of country boys who never had the measles" also "ran through the whole category of complaints that boyhood and babyhood are subjected to." In the new disease environment, officers and medical personnel faced a fundamental tactical problem: how to keep aspiring soldiers healthy enough to get them ready for action. Time proved the only solution. Commanders simply had to wait until the various diseases ran through the available human hosts. For early recruits, like those at Garysburg, it took until early July for the initial epidemics to abate. By then, the troops had begun to move out from the training camps in preparation for the war's first pitched battles.[9]

Believing one military victory might end the war, President Abraham Lincoln pushed his generals to engage the southern armies. In Virginia, Maj. Gen. Irvin McDowell grudgingly put the inexperienced Army of the Potomac in motion in mid-July, hoping to defeat a Confederate force camped near Manassas Junction, thirty miles west of Washington. At the same time, halfway across the country, Gen. Nathaniel Lyon mobilized his Union army of the West to drive away the secessionist Missouri State Guard gathered near Springfield, Missouri. Another 1,000 miles to the southwest, a Union garrison retreated as a large Confederate force approached their isolated fort in the New Mexico desert. These simultaneous campaigns resulted in the battles of Bull Run (July 21) and Wilson's Creek (August 9), and the retreat from Fort Fillmore (July 27), respectively—all of them Confederate victories.

During those first contests, new recruits encountered another serious threat to their health, one that came not from microbes but from the physical environment in which the fighting took place. Even as they fought each other, soldiers on both sides grappled with a common enemy: heat-related illness brought on by a blazing sun. The heat itself was nothing new, of course. Nineteenth-century people were accustomed to being out of doors and dealing with the elements. But the demands of war made prolonged exposure to the sun far more dangerous.

Marching under heavy packs and wearing woolen clothing not conducive to evaporative cooling, Civil War soldiers became especially vulnerable to heat-related ailments. The most common affliction was heat exhaustion, a condition brought on when the body can no longer dissipate enough heat to avoid dehydration and electrolyte abnormalities. As a result, blood pressure drops, causing the victim to faint. The remedy is usually simple. Victims need to cease activity and gradually cool their bodies so that they can again employ the usual defenses against overheating. A second, much more serious condition, known as heat stroke, occurs when prolonged exposure causes the body's methods for regulating temperature to break down. Victims stop sweating and their temperature soars. Multiple vital organs, including the brain, lungs, and kidneys fail and death ensues. Even if one recovers, the damaged internal organs do not function properly, often leaving a survivor permanently disabled.[10]

Inadequate hydration increased the risks of debilitation and death from heat. Today, the USDA dietary reference intakes (DRI) for water consumption recommend that men over the age of eighteen drink at least 3.7 liters (roughly 125 ounces) of fluids per day—more if engaged in strenuous exercise in a warm climate. Given the unreliability of water sources for an army on the move and the limited capacity of army canteens (usually no more than 32 ounces), few Civil War combatants drank their recommended daily allowance of water. The lack of liquids inevitably weakened the body's regulatory systems, especially those designed to cope with prolonged exposure to the sun.[11]

The typical soldier's diet did not help either. The 1860 U.S. Army ration guide called for each man to consume 20 ounces of beef or 12 ounces of pork or bacon; 18 ounces of flour or 20 ounces of cornmeal; 1.6 ounces of rice or 0.64 ounces of beans or 1.5 ounces of dried potatoes; 1.6 ounces of coffee or 0.24 ounces of tea; 0.24 ounces of sugar, 0.54 ounces of salt, and 0.32 ounces of vinegar *each day*. Confederate commissary officers frequently made substitutions—more bacon, less beef, rice instead of beans or potatoes, cornmeal instead of flour—but still tried to adhere to the basic requirements.[12]

If a soldier ate only his daily rations, he consumed between 3,500 and 4,000 calories per day, depending on the type of meat, meal, and starch provided. For many volunteers, that daily ration might have provided more calories than their normal civilian diet. Even so, daily rations lacked suitable quantities of fresh vegetables, a major source of minerals such as calcium, magnesium, and potassium, all of which can be quickly depleted during strenuous activity in extreme heat. Congress authorized adding vegetables to the Union ration in the late summer of 1861, but fresh produce often spoiled before it could be delivered to the front. Dried vegetables proved unpopular, and officers usually did not force their men to eat them.[13]

Regardless of quality, rations were much easier to fill when armies were in camp. Troops quickly learned that, once in motion, an army's available food (and as a result, soldiers' health) declined significantly. Many Union regiments arrived at Bull Run, near Manassas Junction, weak from hunger as supply wagons failed to appear on time. The men of the Thirteenth New York had only crackers and water for dinner and breakfast before the battle. The Second New York entered the fray having not had time to eat any food at all that day. The Eleventh New York had only coffee and hard biscuits for breakfast.[14]

Supply problems proved even worse in the western theater. When General Lyon moved his men into southwestern Missouri to confront secessionist forces, a small Confederate army under Gen. Ben McCullough moved through northern Arkansas aiming to smash Lyon's army and wrest control of the Show-Me State from the Union. As Lyon marched toward Springfield, Missouri, his supply wagons could not keep up and his soldiers had to subsist on one-quarter rations. Similarly, southern soldiers experienced such short rations during their grueling march that hungry men scavenged green corn in the fields along their path. Though it temporarily satisfied their hunger, the unripe corn invited diarrhea that, in turn, led to dehydration and even greater vulnerability to summer temperatures.[15]

Early on, some observers believed that Union soldiers, unaccustomed to the southern climate, might fare worse in summer combat than their Confederate counterparts. During the first months of the war, that prediction looked like prophecy. On the march to Bull Run, one-quarter of the soldiers in the Third Connecticut collapsed in the ninety-degree heat. Several members of the First Minnesota succumbed to what leaders termed "sunstroke." Conditions got so bad that the army's medical director urged Gen. Irvin McDowell not to rush his men to the front in the "hot and sultry" weather, yet dozens still fell to heat exhaustion. The numerically superior northern forces drove the Confederates back in the morning, but the tide turned in the heat of the afternoon. A Confederate counterattack, aided by fresh troops brought by rail to Manassas, forced Union soldiers into a panicked stampede back to Washington.[16]

Excessive heat also helped determine the course of the Battle of Wilson's Creek in southwestern Missouri. Worried about the effects of extreme temperatures, Confederate armies heading toward the battle site began marching at 2 a.m. and halted for the day by noon. Meanwhile, Union soldiers grew weak in the searing sun. A soldier in the First Iowa called the march toward Springfield a "trial of endurance." Dozens of men lay stricken by the roadside, while only twenty-seven of ninety-seven men in his company made it to camp that evening. On the march, journalist Thomas Knox recalled passing "scores of men who had fallen from utter exhaustion. Many were delirious, and begged piteously for water in ever so small a quantity. Several died from excessive heat, and others were for a long time unfit for duty." Some soldiers reported an air temperature of 105 degrees on the day of the battle, and water proved nearly impossible to find. One man, who filled his canteen with nasty liquid from a hog trough, refused an offer of five dollars in gold for the precious substance. Knox saw men so frantically thirsty that they "tore up handfuls of moist earth, and swallowed the few drops of water that could be pressed out." When Union general Nathaniel Lyon fell mortally wounded in the battle, his exhausted army beat a hasty retreat to St. Louis.[17]

Never was dehydration a more serious problem than in the deserts of New Mexico in July 1861. That spring, Confederate authorities developed a plan to conquer New Mexico, Arizona, and eventually take California. If successful, those efforts might provide additional territory for the expansion of slavery and give the South access to Pacific ports. The Confederacy also planned to lay claim to silver mines in New Mexico and gold deposits in California's Sierra Nevada Mountains. Lt. Col. John Baylor led a small group of Texans on the initial military incursion into New Mexico in July 1861. His first target was Fort Fillmore, a Union outpost near the Texas–New Mexico border.[18]

After the opposing forces engaged in a brief skirmish on July 26, the Union commander, Maj. Isaac Lynde, deemed Fort Fillmore untenable and ordered the fort's garrison and civilians to retreat to San Augustin Springs, approximately twenty-five miles away. The Union troops had to march along the Fort Stanton Road, which passed through the southern edge of an infamously hot and arid New Mexico desert known as the Jornada del Muerto (literally, "journey of the dead man"). On July 27, the retreat slowed as the temperature soared to ninety-five degrees. Heat exhaustion quickly dropped many of the soldiers in their tracks. A Union officer recalled that 150 men were "unable to rise or carry their muskets and useless and disorganized in every way." The pursuing Confederate troops under Lt. Col. Baylor encountered numerous stragglers begging for water. "Finding most of them dying of thirst," one wrote, "we gave them the water we had, and were compelled ourselves to go to a spring in the mountain for water." One Confederate soldier wrote, of the desperate, dehydrated boys in blue, "It was beyond anything that it was possible to imagine; many of the men became absolutely insane."[19]

As historian Megan Kate Nelson explains, "Dying of thirst is a terrible ordeal, coming on slowly but steadily in five phases, each phase corresponding to increasing amounts of water depletion." She cites the American naturalist W. J. McGee, who, based on his study of desert rats, distinguished the five phases as "Clamorous, Cotton-mouth, Shriveled tongue, Blood sweat, and Living death." He thought one could only survive the first three phases. Nelson writes, "A person walking in the hot sun without water could progress through these phases in as few as seven hours; the loss of water and sodium through sweating causes blood volume loss and brain cell swelling." Headaches, hallucinations, and seizures plague the desperate victims of death by dehydration. The heat and water depletion proved too much for the Federals, and their diminished force surrendered to Confederates on the evening of July 27.[20]

As those early battles unfolded in the summer of 1861, the once-eager recruits of the Fifth North Carolina took up residence at Yorktown, Virginia. Joining Gen. John Magruder's forces, the men camped near trenches originally dug by British soldiers during the famous Revolutionary War siege eight decades earlier. As at Garysburg a few months earlier, the mere presence of new people on the landscape had important implications for the local environment and for human health. Shortly after arrival the men began to complain of sluggishness, headache, stomach pain, aching joints, fever, and a dry cough, usually accompanied by some combination of nausea, constipation, and diarrhea. By August, evacuation of those still healthy enough to move seemed to be the only remedy. George L. Gordon, a member of the Fifth North Carolina, reported to his wife on August 1 that his regiment was leaving Yorktown the next day: "I believe the object is to get the well men out of this camp where there is a good deal of sickness." Three days later from nearby Warwick Court House, he wrote that only half of his company was present, "the others being left in Yorktown as invalids." By August 31, Gordon noted that "such a large proportion of the Regt are unfit for service, owing to sickness, that those in authority have refused all furloughs to the well." A week later, 78 men in the regiment had died, and Gen. Magruder reported that only 230 out of 1,150 in the unit could muster the strength to take up arms.[21]

Fever, body aches, and gastrointestinal distress are complaints common to a wide variety of ailments. As more and more men fell ill, though, surgeons took note of a rose-colored spotted rash, different from the one associated with measles and a key indicator that the regiment likely suffered from typhoid fever. If that diagnosis was correct, the debilitating symptoms stemmed from the presence of bacteria once known as *Salmonella typhimurium* and *Salmonella paratyphi*, which have recently been reclassified together as *Salmonella enterica*. Those microorganisms invaded the digestive tracts of the soldiers when they ingested contaminated water and food. Undeterred by acid in the human stomach (the salmonellae can stay alive even at a pH of 1.5, roughly twice the acidity of lemon juice) and other cellular defenses, the bacteria multiply for ten to fourteen days and escape into the bloodstream, producing the aches, fever, and digestive problems. Other symptoms, including intestinal bleeding, delirium, and severe stupor, sometimes follow. The most acute cases can result in kidney failure and death. Those who survive begin to show signs of recovery in about four weeks but may remain in a weakened state for much longer.[22]

Infected individuals pass the bacteria back into the environment in their stool, a process that can go on for a year after a person recovers from the disease. The organisms can live for weeks outside the body and are easily transported into streams and groundwater. The salmonellae find their way into the food supply from contaminated water used in cooking, on unwashed hands, or via flies that land on human excrement and then congregate around meal preparation. The Virginia Peninsula, with its permeable soils, countless streams, and a water table that lay close to the surface, offered ideal environmental and geographical conditions for the propagation of the bacteria. Indeed, the disease might have been one of several that killed two-thirds of the original English colonists as they holed up for the winter in their fort at Jamestown. By 1861, though, typhoid fever had become far less prevalent as settlement spread out across the coastal plain and into the Piedmont and mountains. Local communities simply lacked the highly concentrated population necessary to sustain the bacteria.[23]

All of that changed as the Fifth North Carolina and thousands of their comrades in arms converged upon the region. From a biological standpoint, deployment of Confederate forces to Virginia amounted to a massive human migration, one that suddenly offered salmonellae a multitude of potential hosts. Officers knew to locate latrines ("sinks" in military parlance) away from sources of clean drinking water. Leaders also issued orders to police the camps daily to remove trash and filth. But such regulations did little to slow the spread of typhoid fever. Too weak to stumble toward the latrines or unable to reach them in time, sick soldiers often defecated near their tents. As a result, the bacteria flourished in proximity to the cooking fires. When it rained, human waste from the landscape washed into the surrounding streams and invaded the water table. Consumed by sick and healthy alike, tainted food and water perpetuated the insidious ecological cycle. At Jamestown, Confederate troops (like colonists 250 years earlier) became so debilitated that Magruder had to abandon that post by early August. Fully half of the Sixth Georgia Infantry regiment was on the sick list.[24]

Similar scenarios played out wherever soldiers gathered, even in environments less favorable to the bacteria. By December, the disease had weakened garrisons along the Atlantic coast, in the Appalachians, and in various camps throughout Tennessee and Mississippi. Though some soldiers recovered on-site, others had to be sent home for rehabilitation. One southern recruit received a furlough in October 1861 to convalesce at his home near Milliken's Bend, Louisiana. "He has had typhoid fever for a month," his sister wrote, "and has lost forty pounds." Union soldiers ranging from Missouri to the nation's capital also found themselves at the mercy of *Salmonella enterica*. One Union officer in Columbia, Kentucky, wrote to his area commander two days before Christmas to say that a quarter of his regiment had been stricken with typhoid. His soldiers appeared healthy while marching, but once in camp, he reported, "the men sicken and fall like leaves." He begged to attack the enemy: "We would rather die in battle than on a bed of fever."[25]

Civil War surgeons could usually diagnose typhoid from its distinctive rash and the stupor that often enveloped the afflicted. A host of other digestive ailments proved far more difficult to distinguish. Medical personnel tended to lump such diseases together as "diarrhea" or "dysentery," by far the most common illnesses among Civil War soldiers. Modern physicians recognize the two as separate problems. Technically, diarrhea is a symptom, not a disease, and is common to a variety of ailments. Dysentery usually manifests itself as diarrhea with mucus and/or blood in the stool. Though it can result from amoebic parasites, dysentery more often stems from bacteria, including *Shigella*, *E. coli*, and other forms of salmonella. All of those microbes flourished among human populations set in motion by the war. As medical historian Margaret Humphreys writes, "The Civil war soldier lived amid a soup of fecal organisms," and every recruit—Union and Confederate—likely suffered from bowel maladies at some point in his service. Most soldiers suffered on multiple occasions. Over the course of the war, Union medical personnel recorded nearly 1.75 million cases of diarrhea or dysentery, which resulted in 45,000 soldier deaths. Confederate statistics are not nearly as comprehensive, due to the loss or destruction of records, but southerners probably suffered at similar or even higher rates. Perhaps Walt Whitman, a nurse in Union hospitals, put it best when he soberly observed that the "war business is about nine hundred and ninety-nine parts diarrhea to one part glory."[26]

The booming bacteria populations in the camps make it difficult to determine the exact microbial agent responsible for recorded bowel complaints. Milder cases of dysentery likely stemmed from one of several species of *Shigella*, an organism that lives in human stool. Ingesting even a tiny portion (such as a particle of feces too small to see) can be enough to infect an otherwise healthy person. *Shigella* often passes from one person to another via contaminated food or water and can also be spread by flies. Symptoms, sometimes including bloody diarrhea, appear within one to two days. In generally healthy individuals, the worst symptoms disappear in about a week and the individual acquires immunity to that specific strain of *Shigella* but remains susceptible to other forms of the bacteria. *E. coli* and salmonella could trigger more severe bouts of dysentery. Many strains of *E. coli* inhabit the intestines of humans and other warm-blooded animals. Most varieties pose no threat of disease and actually aid digestion, but some can be agents of dysentery, either by themselves or via production of a toxin that causes intestinal distress. Incubation takes a bit longer than for *Shigella*, and the resultant cramps, diarrhea, and fever are usually worse and longer-lived, sometimes persisting for several weeks. In the military camps, where the microorganisms proliferated, a Civil War soldier could easily be infected with different strains over the course of his military service. A host of viruses can also cause bowel problems. Together these microorganisms accounted for much recurrent diarrhea among troops on both sides.[27]

In addition to generally crowded camp conditions, the dysentery-causing bacteria also benefited from the soldiers' lack of experience with cleanliness and food preparation. In patriarchal nineteenth-century society, women usually took care of household cleaning and cooking. According to Union surgeon-in-chief Charles Tripler, "The individual man at home finds his meals cooked and punctually served, his bed made, his quarters policed and ventilated, his clothing washed and kept in order without any agency of his own, and without his ever having bestowed a thought on the matter." Even officers entrusted with bringing cleanliness and hygiene to the camps

The Seventh New York at camp near Washington, D.C. Forced to clean and cook for themselves, soldiers frequently became sick from their unhygienic practices. From Johnson and Buel, *Battles and Leaders of the Civil War*, 1:159.

often lacked those skills. As Tripler put it, "In ninety-nine cases in a hundred [the commander] has given no more reflection than the private to these important subjects."[28]

Thorough cooking could render infected food harmless, but many soldiers did not know either the techniques or cooking time required to kill the bacteria. Basic handwashing might also have curbed dysentery (as it does today), but in the general filth of the camps, the water in which men bathed might be as contaminated as the food. One soldier in the Thirty-Fifth New York wrote home that sickness had run rampant through his regiment, but he tellingly noted that "more than one half of it is caused by the carelessness of the men." The same problem plagued Confederate armies; Robert E. Lee lamented that soldiers "bring [sickness] on themselves by not doing what they are told. They are worse than children, for the latter can be forced." Though Lee's comments might be generally correct, some soldiers did adopt practices that aided in the struggle against various ailments. Kathryn Shively Meier argues that soldiers frequently practiced "self-care," such as taking baths, washing their clothes, scavenging fresh food from local farms, and even temporarily leaving their units, or "straggling," to rest and find shelter from the elements. Such measures might have eased individual symptoms, but they did little to alleviate the problems in the camps.[29]

In the constant fight against diarrhea and dysentery, army doctors also used any remedy that seemed viable. One surgeon claimed that he had cured some soldiers by feeding them raw Irish potatoes doused in vinegar. Another swore that eating large quantities of roasted peanuts cured diarrhea, though a counterpart declared them "utterly worthless" as a treatment. Ironically, conventional medical wisdom at the time suggested that eating fresh fruits and vegetables actually caused diarrhea. Experience on the battlefield soon shattered that notion, as medical personnel came to understand that "when fresh vegetables were issued to the troops, or obtained by foraging, the number of soldiers on sick report with fluxes promptly diminished."[30]

Doctors prescribed a wide variety of elixirs for bowel complaints, though the medicines were hardly worthy of the name. Generally consisting of a dangerous mixture of unhealthy chemicals — frequently derived from opium or mercury — the concoctions often compromised the physicians' Hippocratic oath: "First, do no harm." William H. Taylor, a Confederate surgeon in Charlottesville, Virginia, described his treatment program: "In one pocket of my trousers I had a ball of blue mass [a mercury compound], in another a ball of opium. All complainants were asked the same question, 'How are your bowels?' If they were open, I administered a plug of opium; if they were shut I gave a plug of blue mass."[31]

Contemplating either treatment is enough to make modern physicians shudder. To treat diarrhea, some army doctors ordered that opium tinctures, often in combination with other drugs, be injected directly into a soldier's sore rectum. The opium momentarily blocked the pain, but understandably most patients preferred to ingest the narcotic in more traditional ways. Blue mass, a chemical compound commonly called mercurous chloride, or calomel, might be an effective purgative (it had been an accepted medical treatment since the sixteenth century), but the side effects could be lethal. Taken internally, the compound sometimes broke down into two highly toxic substances: mercury and mercuric chloride. During the war, as Roy Morris Jr. has written, "some patients were given a dram of calomel every hour, causing their faces to swell, their tongues to jut out of their mouths, and their saliva to gush forth at the rate of anywhere from a pint to a quart every twenty-four hours." Other patients had their teeth and hair fall out, or developed mercurial gangrene, in which the soft tissue of the mouth rotted and fell out.

Soldiers suffered from a variety of ailments. Disease was responsible for two out of every three Civil War deaths. From *Harper's Weekly*, March 11, 1865.

The surgeon general banned the use of calomel in May 1863, but many doctors continued prescribing it for many complaints because they believed it worked. If physicians employed it less frequently in southern armies, it was only because the Union naval blockade eventually rendered all medicines in short supply.[32]

While military life in wartime exponentially increased common illnesses that came with diarrhea, it also created nearly ideal conditions for the spread of venereal diseases such as syphilis and gonorrhea. The bacteria that cause both diseases are highly sensitive to increased oxygen and temperature variations, meaning that they do not survive long outside the human body and people are the only carriers. Human beings in motion and changes in sexual behavior—both common in wartime—are crucial to creating biological conditions that allow the microorganisms to proliferate. Tracking sexually transmitted diseases is difficult enough in the modern world; charting the spread of syphilis and gonorrhea with any precision during the Civil War is even more problematic. Even so, medical personnel on both sides were familiar with the diseases, allowing at least a glimpse into that new microbial environment.³³

In the prewar North, syphilis occurred most frequently among single men in urban industrial areas along major transportation routes. Boston, Milwaukee and its environs, and Louisville (the nation's twelfth-largest city in 1860) all seemed to have high rates of infection. With the onset of war, even married Union soldiers effectively lived as single men within a diverse and mobile population, creating similar conditions to cities even in smaller towns and rural areas. The only missing ecological element was a reservoir of potentially infectious bacteria, and the numerous prostitutes who plied their trade in and around the army camps unwittingly provided this. As early as May 1861, a Michigan soldier stationed in Washington, D.C., noted, "If the men pursue the enemy as vigorously as they do the whores they will make very efficient soldiers." During the war, Washington's prostitute population increased more than tenfold to nearly 7,000 women working out of 450 brothels. The number of prostitutes in Nashville grew from 200 in 1860 to more than 1,500 by 1863, when the Union-occupied city served as a major supply base for Federal armies.³⁴

Though we lack comparable data about the prevalence of venereal disease in the antebellum South, the Confederacy experienced a similar boom in prostitution when hostilities began. In Richmond, the arrival of Confederate soldiers provided thousands of potential new clients for the city's working girls. With space for brothels at a premium due to the influx of troops, one bawdy house opened across the street from a soldier's hospital run by the Young Men's Christian Association. Women called to the men from the establishment's windows, prompting the hospital director to complain to the provost marshal that the scantily clad temptresses inhibited recovery of the sick soldiers. By the summer of 1861, Confederate authorities in Richmond reported hundreds of new cases of syphilis and gonorrhea. Venereal disease quickly came to rival measles as an early scourge of Confederate camps. When Col. William Barksdale arrived in the capital with his Mississippi Regiment in June 1861, he wrote that his regiment was beset with sickness, primarily from measles and promiscuity. A month later, the Tenth Alabama reported that 68 of its 1,063 men had been infected with gonorrhea.³⁵

Treatment for venereal diseases usually involved prescription of mercurial compounds. Physicians were so quick to prescribe the drug that it became part of an informal prevention campaign. Nearly every Civil War soldier on both sides heard the quip "A night with Venus, a lifetime with Mercury." Doctors also tried pokeroots, sarsaparilla, sassafras, jessamine, prickly ash, silkweed root in whiskey, pine rosin pills, and various other substances injected directly into the urethra. Nothing, however, cured the diseases, and they usually flared repeatedly in later years.[36]

In addition to treating afflicted soldiers, Union authorities tried on several occasions to remove sources of new infection. In 1863, in Nashville, authorities rounded up as many of the city's prostitutes as possible, loaded them onto a military transport vessel, and shipped them north. After several cities refused to take them, the women eventually returned to Tennessee's capital. When relocation failed, military officials tried to regulate and restrict the sex trade. Surgeons issued licenses to prostitutes and required them to undergo weekly health inspections. Those who showed signs of gonorrhea or syphilis had to stop plying their trade and report to a special hospital for treatment. Authorities arrested any woman working without a license. The system worked well enough that other occupied cities adopted similar measures to prevent the spread of this predictable curse of soldiers.[37]

In 1861, surgeons on both sides feared that soldiers might have to cope with another common scourge of war: cholera. Caused by the bacterium *Vibrio cholerae*, the disease enters the body through the consumption of contaminated food or water; flies can also spread the bacterium to food. Cholera, which thrives in temperate, brackish water, appears where large populations live in unsanitary conditions or do not have adequate clean water. Violent purging of watery vomit and excessive watery diarrhea are typical first symptoms of the sickness. Patients can evacuate up to five liters of fluid a day, which leads to severe dehydration, organ failure, and death.[38]

Though cholera does not spread through direct human contact, it is a disease that "travels" or hops from one site to another. Once introduced into a stable water (or less commonly, food) supply, the bacteria can multiply and produce an epidemic. From there, it typically requires a human host or some other carrier to move to another locale and establish itself amid another dense population. In the United States, cholera most often surfaced in large cities or among other highly concentrated populations such as those who made their way west along established trails. Its first American appearance might have come in 1832, when it established itself in New York City and moved west and south to other urban areas. During 1849, the disease killed former president James K. Polk as well as 3,000 people in New Orleans, 5,000 people in St. Louis, 15,000 in New York City, and thousands more who were en route to California in search of gold. Between the two epidemics, an estimated 150,000 Americans died, while thousands more suffered in various smaller outbreaks until the eve of the Civil War.[39]

Outside the United States, armies seemed particularly vulnerable. During the Crimean War from 1853 to 1856, French, British, and Russian armies lost tens of thousands of soldiers to cholera in ports along the Black Sea. Conditions in Civil War camps favored the disease, but cholera never became a major problem for American soldiers. Perhaps because severe epidemics had run their courses just before the war and those who survived had developed some immunity, the disease was not able to travel so easily into and between Civil War armies. It did break out again immediately after the war and moved along newly established railroad routes in the West, but during the war, it seems to have been unable to revive itself and settle in among the troops.[40]

For the most part, Civil War soldiers also avoided typhus, another deadly disease that had wreaked havoc among armies across the world for 500 years. The most common and deadly variety in wartime is known as "epidemic typhus," because it runs rampant through a vulnerable human population, much like measles or other viral infections. Epidemic typhus, though, is not viral. It is caused by a bacterium called *Rikettsia prowazekkii*, an organism found throughout the world. It does not spread through simple contact or via infected food or water and should not be confused with typhoid. Instead, typhus requires arthropod vectors, usually insects such as body lice, to move the bacteria from one person to another. Simply stated, a louse bites an infected human and the microorganism takes up residence in the arthropod's alimentary tract. Some infected lice move to another host and leave behind feces when they eat. When the victim scratches the bite the bacteria-laden excrement enters the bloodstream through the break in the skin. People living in close quarters in unsanitary conditions are especially vulnerable to lice-borne disease. Malnutrition or poor dietary habits also enhance the chances of infection. Indeed, epidemic typhus is commonly called "ship," "camp," or "jail" fever because it so often proliferates in those settings.[41]

Chronically malnourished, often fatigued, and crowded into unsanitary army camps, Civil War soldiers would appear especially vulnerable to epidemic typhus. By all accounts, the necessary vectors were not in short supply. Commonly called "graybacks," body lice proliferated wherever armies gathered. As one soldier joked, the bugs "are in for the duration of the war, and they never desert their command." Then, as now, lice infection carried a social stigma and, among new recruits, was initially a source of shame. As they settled into the routines of army life, however, most soldiers simply accepted lice as one of the consequences of war. One veteran quipped that he could not sleep without "a few graybacks gnawing on him." Whenever they had downtime, men deloused as best they could, often describing the process with military terminology. Discarding infected clothing came to be known as "giving the vermin a parole" and turning one's clothes inside out constituted a "flanking movement." To relieve the boredom of camp life, soldiers sometimes held "lice races" and placed bets on their favorite insects. Stories also circulated about how quickly camp lice learned the habits of their human hosts. The troublesome insects "knew all the bugle calls," one man recalled, "and had become so expert in drill as to go through the battalion movement quite accurately, and to have their regular guard mountings and dress parades." Small wonder that the first use of the word "lousy" can be traced to the Civil War.[42]

The bacteria responsible for typhus were also readily available, perhaps within the American population and certainly among recent European immigrants to the United States. Yet with vectors, bacteria, and a potentially vulnerable population all in place, typhus never became a problem during the Civil War. Save for an unconfirmed outbreak in Wilmington, North Carolina, near the end of hostilities, no epidemics have been positively identified. Union records reported 2,624 incidents of typhus and 958 deaths, but even at the time, many physicians believed that those cases might have been misdiagnosed.[43]

Why were American soldiers able to avoid the disease? No one can say for sure, but epidemic typhus also declined in Europe during the nineteenth century, only to return during World War I. Maybe the Civil War occurred during a natural lull in the worldwide rate of infection. Or perhaps Americans had some sort of acquired immunity from having contracted other, milder forms of typhus or related diseases. Given the overall susceptibility of Civil War soldiers to epidemics, such explanations seem unlikely. To date, the most plausible theory, suggested by medical historian Margaret Humphreys and backed by at least some clinical research, holds that the American body louse might have been a "less efficient disease vector than its European counterpart." Currently, entomologists make no distinction between European and American body lice, but it may be that some tiny ecological difference in the American variety rendered it less capable of transmitting *Rikettsia prowazekkii*. Since typhus also kills the infected louse, any disease-carrying European lice hitchhiking into the country might well have died off before they had a chance to establish a population and spread the infection. Evidence to support that theory remains sparse, but at least one study showed a much higher rate of infection among guinea pigs exposed to European lice than in those exposed to American lice.[44]

When it came to other indigenous insects, Civil War soldiers were not so lucky. During the warm months, troops on both sides had to fend off clouds of mosquitoes that descended on encampments and followed soldiers into battle. Though neither the recruits nor their surgeons knew it, some of those winged pests carried parasites that caused malaria, a disease that rendered thousands of men unfit for battle, especially during the first two years of the war. Exactly how and when malaria became established in the United States remains uncertain. Some evidence suggests that anopheline mosquitoes capable of carrying malarial parasites were native to North America. The parasites themselves, though, most likely came from Europe and Africa, transported in the blood of European colonists, slaves, and slave traders. By the mid-1600s a comparatively milder form of malaria (known as *Plasmodium vivax* and endemic in parts of Europe) had already become a problem in colonial Virginia. By the 1680s, increased importation of slave labor brought a much more virulent and potentially deadly variety (known as *Plasmodium falciparum* and likely native to Africa) into the South. Because the plasmodium that causes *vivax* malaria is more tolerant of cooler conditions, it followed settlers into the Midwest and even more northern climes. Requiring warmer weather, *falciparum*, or "pernicious" malaria, became endemic in the coastal and Deep South. By 1860, military officers and surgeons knew the disease well, though they remained ignorant of its cause or how it spread. Like doctors elsewhere, they associated it with putrid air or vapors emanating from swamps, rotting vegetation, or stagnant water. Indeed, the term "malaria" derived from Italian words for "bad air."[45]

Malaria has a complicated ecology, but essentially it requires three things to sustain itself among people: parasites, anopheline mosquitoes, and nonimmune human hosts. A female *Anopheles quadrimaculatus*, the most common carrier in the United States, bites an infected victim, imbibing the *Plasmodium*. In search of blood meals to facilitate breeding, the now parasite-carrying mosquito bites an uninfected person and injects the parasites into the bloodstream. From there, the *Plasmodia* move to the liver where they multiply, alter their form, and attack cells within that organ. Infected liver cells swell and burst, releasing large numbers of parasites back into the bloodstream where they, in turn, infect red blood cells. As the affected cells rupture, the victim experiences the telltale fever spikes, chills, sweating, and nausea associated with malaria. In time, as the body marshals its defenses against the infection, the parasites retreat to the liver, where they multiply and start the cycle over again. This pattern of acute symptoms, recovery, and relapse can continue for years if the disease remains untreated. Though the parasites — especially the more virulent *P. falciparum* — can kill, malaria more often debilitates a victim and lowers resistance to more serious and deadly secondary infections such as pneumonia.[46]

Some of those who battle and survive malaria develop partial immunity. In such cases, reinfection will produce few or no symptoms, though such protection works only against that specific strain of *Plasmodium*. Mothers can pass this acquired immunity to infants during pregnancy, protecting children early in life. Children who survive milder cases of infection in their first years develop the same semi-immunity. Where malaria is prevalent, human populations can, over time, also develop genetic traits that help protect red blood cells. The best known of these is sickle-cell trait (not to be confused with sickle-cell anemia), but researchers have also identified several other blood anomalies that offer protection. As malaria establishes itself in a relatively stable human population, parasites, vectors, and hosts move toward uneasy equilibrium. Victims relapse and recover, allowing the parasites to survive; new nonimmune hosts appear periodically; and the mosquitoes find adequate blood meals to facilitate breeding. The disease becomes chronic and seasonal with predictable outbreaks of variable intensity occurring during summer and early autumn when *Anopheles quadrimaculatus* are active.[47]

Because anopheline mosquitoes need fresh, slow-moving water in which to lay their eggs, the local environment plays an important role in sustaining malaria. The vectors prefer freshwater with rooted or floating vegetation that conceals larvae from predators. Swamps, ponds, and sluggish streams all provide suitable habitats, as do rain barrels, drainage, and irrigation ditches, rutted roads, and other trappings of human life in which water and detritus accumulate. Generally speaking, malaria and the mosquitoes that spread it are more prevalent during early human settlement. As people clear forests, drain swampy terrain, and move away from mosquito-infested areas, anopheles populations decline, another factor that makes for the predictable seasonal arrival of malaria in long-occupied regions where the disease has become endemic. Though Civil War surgeons had at best a limited understanding of malaria, their belief that swamps and rotting vegetation facilitated its spread was not far off the mark.[48]

Since the South had more than its share of swampy habitat, prognosticators on both sides believed that the southern environment might, in effect, be weaponized and turned against the Union. According to some observers, "Unacclimated soldiers would quickly be disabled by the pernicious malarial fevers that prevailed over more than half the region, and for nearly half the year, south of the Ohio [River]." From the first, northern officers tried various deterrents thought to be effective against the disease. In September 1861, Gen. George B. McClellan issued standing orders that soldiers in the Army of the Potomac be given hot coffee immediately after roll call "as a preventive of the effects of malaria." Such efforts, of course, proved futile.[49]

Where malaria had become endemic, the machinations of war scrambled established relationships among parasites, vectors, and hosts. Union officers moved troops along southern waterways and coastlines, bringing nonimmune men into prime anopheline habitat. Wagons and artillery left ruts that filled with rainwater. Soldiers constructed trench lines, ditches, and latrines, unwittingly providing more breeding sites for *Anopheles quadrimaculatus*. Recruits from the Ohio Valley and other temperate northern regions sometimes brought *P. vivax* into Union camps. Freed slaves, some of whom enjoyed acquired or genetic immunity to malaria, inadvertently furnished another source of potential infection for their Yankee liberators. It did not take long for the effects of such actions to show. After a Union force captured the islands of Port Royal Sound in South Carolina in the late autumn of 1861, an officer lamented that the "deadly malaria that rises from the swamps" had sickened hundreds of the occupying Union soldiers.[50]

Though malaria might not kill, it could debilitate a victim for years. As one scholar explains it, the spiking fever, chills, nausea, and dehydration "severely undermined an individual's health, causing malnutrition, cerebral anemia, and cognitive impairment." Symptoms could recur daily, or once every two, three, or four days, prompting doctors to describe it as intermittent, quotidian, tertian, or quartan fever. Soldiers battling the disease became more susceptible to respiratory infections or death from dysentery and diarrhea. As the war progressed and Union armies advanced deep into southern territory, malaria became common in northern ranks. By the time hostilities ceased, approximately 1.3 million Union soldiers had contracted the disease.[51]

Even so, the South's malarious environs were not enough to turn back northern troops, in part because physicians had an effective treatment and preventative for malaria: quinine. Quinine is derived from the bark of some forty species of trees in the cinchona genus — tropical plants that grow in the western Amazon region along the eastern slopes of the Andes Mountains in South America. It had a bitter taste, which discouraged soldiers from taking it unless they dissolved it in whiskey. Because it proved highly effective at reducing fever and relieving pain, doctors began prescribing it for almost

Quinine proved effective against malaria, but because of the medicine's bitter taste, doctors usually administered it with whiskey. From *Harper's Weekly*, March 11, 1865.

any illness. One Massachusetts soldier declared, "Quinine is used indiscriminately at the commencement, middle, and end of every disease." The drug became so popular that by the end of the war Union soldiers had ingested more than a million ounces of quinine and other derivatives of the cinchona plant. Off and on during the war, some soldiers, especially on the Union side, slept under mosquito netting to increase comfort, thereby stumbling on an effective prophylactic for malaria. But the mesh was often expensive and in short supply. Not until the latter stages of the conflict, when Union hospitals began to use "mosquito bars," as the netting was called, did the practice get a more widespread trial.[52]

Confederate soldiers, too, fell victim to malaria, though incomplete records make it impossible to judge the extent of infection within their ranks. What we do know is that southern demand for quinine exploded during the first years of the war. Because of the North's naval blockade, supplies of the precious medicine all but dried up as the conflict wore on. As historian Andrew McIlwaine Bell has shown, the South's continuing need for quinine and, by implication, the extent of malarial infection among Confederate troops, can be charted in the soaring prices commanded by the elixir. An ounce of quinine sold for $4 in the summer of 1861. It fetched $22.50 in 1863 and quickly became a favorite commodity for smugglers and blockade-runners. Two years later, those who could afford quinine were paying $600 per ounce, a 15,000 percent increase! Even allowing for near-worthless Confederate currency and the actions of war profiteers, those figures suggest that southerners, too, quickly became locked into a war against invading soldiers and indigenous parasites. Indeed, the North was able to deal with the ecological disruption created by war and turn it to its advantage. The relatively easy access to quinine and the North's ability to keep the medicine out of southern hands might well rank as one of the Union's most successful military strategies.[53]

Malaria was not the only mosquito-borne disease that southerners thought might protect them from conquest. Those who speculated about environmental hazards also touted the "yellow fever zone," noting that this "scourge of the tropics" would easily "decimate any northern armies that might penetrate the Cotton States." Such predictions were well founded. Yellow fever results from a virus carried by a different species of mosquito, the female *Aedes aegypti*. While the *Aedes* requires freshwater breeding sites, it prefers small bodies of standing water like rain barrels, mud puddles, and other reservoirs and containers common around cities. Consequently, in the United States, yellow fever is almost invariably an urban disease. Like the anopheline carriers of malaria, *Aedes* spreads yellow fever by injecting the virus into the human bloodstream. The resulting infection can appear in a mild form that resembles influenza and from which a victim usually recovers in several days. But it can also produce far more serious symptoms, including high fever, chills, and jaundice (hence the name). In such cases, the virus attacks the kidneys, liver, and immune system, bursting blood vessels and causing massive internal bleeding as organs fail. Some patients still recover and gain immunity, but others lapse into a coma and die. A severe case of classic yellow fever often proved easy to identify—an infected person vomited a vile black substance that resembled coffee grounds but was actually coagulated blood. This "black vomit" usually signaled imminent death. Because yellow fever can take several forms, mortality rates are difficult to determine with precision. Even the best estimates suggest that mortality occurs in 10 to 60 percent of victims.[54]

Aedes is extremely sensitive to cooler temperatures and cannot survive winter even in more temperate parts of North America. As a result, yellow fever nearly always appeared as a summer epidemic that wreaked its deadly havoc until the first killing frost. Yellow fever, like cholera, was a traveling disease, one that often depended on fast-moving sailing vessels to transport the contagion (and sometimes vectors) from tropical regions into American ports. While at sea, an infected ship frequently flew a yellow flag. The color yellow is code for the letter "Q," indicating quarantine. In time, "yellow jack" (a reference to the flag) became slang for the disease itself.[55]

Though unaware of what caused yellow fever, Civil War surgeons knew it had a reputation for decimating southern port towns. New Orleans had been especially vulnerable; the disease killed nearly 10,000 people there in 1853. When Union forces captured the Crescent City in the spring of 1862, some residents believed that a summer epidemic might allow the Confederacy to reclaim the port. As the locals put it, yellow jack "will likely soon make its appearance . . . and give brief dispatch to the Northern soldiers who have dared to invade its domain."[56]

That potentially liberating epidemic never materialized, thanks largely to the efforts of Gen. Benjamin F. Butler, the Union officer in charge of the occupied city. Thinking that yellow fever resulted from generally unsanitary conditions, Butler put citizens to work scouring the city's streets, ditches, and drains. He also issued a strict quarantine on any ship arriving from an infected port or whose passengers showed signs of yellow fever. Those vessels had to anchor seventy miles away for thirty days or until the shipboard epidemic subsided. Such methods proved highly effective. Cleaning streets and clearing drains limited standing water in the city and reduced breeding sites available to the mosquito vector. Though a few infected persons inevitably slipped through, the quarantine effectively prevented an epidemic. In later years, Butler remembered one sick passenger who arrived on a steamer from Nassau and somehow escaped quarantine. As the general put it, "The city being clean and the atmosphere pure, the fever did not spread, but died out with the victim." Allowing for the state of medical knowledge at the time, Butler's assessment offers a near perfect ecological explanation of how and why occupied New Orleans avoided the ravages of yellow jack. Indeed, over the course of the entire war, the disease claimed only eleven victims in the Crescent City.[57]

Like malaria, yellow fever did not drive northern armies out of the South, though the disease continued to take a toll in some coastal towns. The worst epidemic occurred in Wilmington in 1862. New Bern and Beaufort, North Carolina, as well as Charleston and Hilton Head, South Carolina, reported severe episodes of infection during the war. Yet, as Andrew McIlwaine Bell argues, whether real or imagined, the terror of yellow jack could prove a powerful deterrent to Yankee plans. In 1862, alone, fear of exposing northern troops to the dreaded affliction helped prevent the Union from launching operations against posts in Texas and South Carolina as well as Wilmington, where the Federal garrison knew all too well the fever, chills,

Cartoon celebrating Benjamin Butler's military occupation of New Orleans in 1862. His efforts to clean up the city probably helped reduce the mosquito population and impede the spread of yellow fever. From *Harper's Weekly*, January 17, 1863.

black vomit, delirium, and death associated with one of the South's worst diseases.[58]

Perhaps because yellow fever conjured such fear among northern leaders, some southerners never gave up on using the disease as a biological weapon. In 1864, with the war going badly for the Confederacy, a Kentucky doctor named Luke Pryor Blackburn hatched an elaborate plot to in-

Sickness

33

fect northern and Yankee-occupied southern cities with the deadly disease. Like other physicians at the time, Blackburn did not know how yellow fever spread. He thought it might pass to others via vomit, blood, sweat, and other bodily secretions from infected victims. With help from two other Confederate sympathizers, one of whom offered seed money, Blackburn surreptitiously retrieved blood- and vomit-soaked clothing from some yellow fever patients he treated in Bermuda. Placing the garments in trunks, the good doctor sent one of his coconspirators to auction off the clothes (and the trunks) in several cities, including Washington, D.C. According to some sources, Blackburn even prepared a special chest of "infected" clothing to be delivered to President Lincoln. While waiting for the distribution of the infected garments, Blackburn took refuge in Canada.[59]

The plan fell apart when one of the coconspirators, who had been promised $100,000 for his efforts, only received $50. He confessed to the Canadian authorities on April 12, 1865. By then Lee had surrendered. Two days later, Lincoln fell to John Wilkes Booth's bullet. (U.S. agents also believed for a time that the doctor might have been involved in Booth's assassination plot.) The Canadian government rushed to arrest Blackburn for "plotting an act of war against the United States," but Blackburn eventually beat the charge in court. It did not hurt that the infamous trunks had since disappeared and that the coconspirator's account could not be verified. Given the need for mosquito vectors and a host of other ecological factors, Blackburn's plot had no chance of success. But accurate information about the spread of yellow fever did not become available until 1897, ten years after the doctor's death. Undeterred by the acquittal, some in the press vilified the Kentucky doctor, calling him "Dr. Black Vomit" or "Yellow Fever Fiend." Those appellations surfaced again when Blackburn ran for the Kentucky governorship in 1879, a post that he eventually won. His work as governor, including efforts to reform the state's prisons, helped to rehabilitate his reputation, though tales of his biological campaign against the North followed him for the rest of his days.[60]

Just as Union soldiers feared the South's tropical diseases, southern commanders worried about the worst scourge of northern cities: smallpox. Like measles, smallpox stems from a virus that spreads through human contact. Anyone infected could expect at least a month-long life-threatening bout with headache, fever, chills, severe backache, and vomiting. Victims also developed a rash that evolved into oozing pus-laden sores that filled a patient's room with stinking vapors. Worse, the virus, now known as *Variola major*, is a hardy microbe that can survive exposure to air and live outside the body for weeks. Even when the afflicted recovered, particles from the crusty scabs that covered the sores might leave bed linen or clothing laden with live virus capable of infecting others. The death rate during a given epidemic depended on a host of factors, including a victim's age, the quality of medical care and nutrition, and the particular strain of *Variola*. Even if the virus itself was not lethal, it so weakened the body that pneumonia or another secondary infection might move in and kill. In previously unexposed populations, the death rate from smallpox might be as high as 90 percent; among younger people in regions where the disease was not new, it might be as low as 8 percent. According to the Centers for Disease Control, "Historically, *Variola major* has an overall fatality rate of 30 percent." Survival was no guarantee of a return to good health. Smallpox could leave victims with horrible disfiguring scars or a host of other medical complications, including blindness.[61]

When the Civil War began, *Variola major* already had a long legacy of destruction in North America. It had played a central role in the decline of Native American populations, and a devastating epidemic during the Revolutionary War had not only killed and disabled soldiers but also profoundly affected military strategy. Out of that experience, Americans had developed three main methods for coping with smallpox, even though doctors did not understand its cause. The first and simplest tactic was avoidance, by either quarantining the afflicted or fleeing in the face of an epidemic. A second defense, developed in the Americas in the early eighteenth century, is today known as inoculation or variolation. That process involved placing lymph or other infected fluid from a smallpox victim into a healthy person, usually via an incision. It was a dangerous procedure. One could die from complications associated with inoculation or from smallpox itself, but if one survived—sometimes with a milder infection—immunity followed. A third measure, vaccination, pioneered by English doctor Edward Jenner in 1796, made use of cowpox, a related but much less severe disease. Given to a healthy person, cowpox (again barring complications) could provide immunity to the more lethal *Variola major*.[62]

By 1860, vaccination had become more common in northern cities than in rural parts of the South and Midwest. In the summer of 1861, Union surgeon Charles Tripler immediately vaccinated all the soldiers in Washington and ordered all recruits to be vaccinated before their units left home. In December 1861, he had brigade surgeons vaccinate all the soldiers who managed to evade it in the summer. The efforts paid dividends as no major smallpox epidemic broke out in the Army of the Potomac.[63]

Nevertheless, the specter of smallpox terrified Union troops, even if they had been vaccinated. In more remote parts of the western theater, far from the front and lacking access to well-trained doctors and proper vaccines drawn from cowpox, some soldiers sought to protect themselves against the contagion. More than a few troops attempted inoculation or vaccination "arm to arm." Using whatever sharp object happened to be on hand, outwardly healthy soldiers took pus (or anything else that oozed) from pustules on infected comrades and placed it in cuts or other openings in their own unafflicted skin. The results of this form of self-care could be catastrophic, as it invited infection or worse, a raging case of smallpox. Moreover, to the untrained soldier's eye, lesions from syphilis and several other diseases might look a lot like *Variola major*. Though the exact extent of noncarnal infection with syphilis remains unclear, at least one documented instance among an Illinois contingent left some of those men facing a lifetime of mercury without the pleasures of a night with Venus. Given the overwhelming fear of smallpox, that scenario might well have been repeated in other Federal camps.[64]

Although the Union had its share of vulnerable recruits, southern leaders knew that far fewer Confederate troops had been vaccinated and that the disease was much less common in the South. Because *Variola* could survive exposure to air and be spread via infected clothing, sheets, and blankets, Rebel officers constantly worried that it might be used as a biological weapon. British officers had tried to infect certain Native American enemies during the Seven Years' War, and American colonists believed that England intended to employ similar tactics during the Revolutionary War. In the spring of 1862, some in the South thought their worst fears had been realized.[65]

In late May 1862, Col. Andrew Jackson Jr., deputy commander at Fort Pillow on the Mississippi River in Tennessee, agreed to an exchange of prisoners with a Federal commander at Alton, Illinois. Two or three of the 202 prisoners sent to Fort Pillow showed obvious symptoms of smallpox. When Gen. John Bordenave Villepigue, the fort's commander, returned and learned of the problem, he expressed his displeasure but stopped short of accusing the Union of intentional infection. "I do not presume that you are in any way responsible for so barbarous an act," he wrote to his Union counterpart, but "I demand that your government disown the act by receiving these prisoners back into its lines and caring for them until every symptom of the infection has disappeared from their midst." Confederate general P. G. T. Beauregard, the commander of the western theater, was equally circumspect. While refusing to blame any specific individual, he made it clear that "to send us prisoners afflicted with contagious diseases of a dangerous and deadly character, is in my judgment, violative of all ideas of fairness and justice, as well as of humanity." From all indications, Fort Pillow did not become infected as a result of the attempted exchange. Indeed the virus might have remained mostly absent from Confederate ranks until the late summer of 1862, when troops moved into Maryland. At some point after *Variola* made its appearance in their ranks, southern soldiers also began to experiment with arm-to-arm "vaccinations," with the same disastrous results. Meanwhile, the disease continued to ravage the Union prison at Alton, so much so that officials began to quarantine infected Confederate prisoners on islands out in the Mississippi River away from the main facility. Those who succumbed to the disease were also buried there. One of those locations eventually acquired the name "Smallpox Island."[66]

By the spring of 1862, the war had already created a new and distinct pattern of microbial exchange. Generally speaking, highly contagious viral diseases associated with the North, such as measles and smallpox moved farther into the South and Midwest. The most lethal southern maladies, malaria and yellow fever, both tropical and spread by mosquitoes, required that nonimmune hosts come to them, something the Union unwittingly facilitated with its invasion of the Confederacy. Though cooler temperatures in the North limited malaria's spread, it migrated into the region in the bodies of those who returned after the war. Those who contracted yellow fever might not come home alive, but rudimentary efforts to improve sanitation and public health kept the deadly contagion from doing its worst among Union troops. Meanwhile, military encampments on both sides turned formerly rural areas into instant cities where crowded conditions and the sexual habits of soldiers created nearly ideal conditions for propagation of dangerous bacteria. Most men avoided typhus and cholera, but typhoid, dysentery, diarrhea, and venereal disease ran rampant. Those patterns held for the rest of the war, with periodic fluctuations as troop movements and the arrival of fresh recruits again jumbled the relationships between microbes and hosts. Movement of freed slaves during and after the war would also provide new human hosts for various microbes set loose during the fighting and became one of the war's enduring environmental legacies.[67]

For those soldiers who survived the camp epidemics and remained healthy enough for combat, the battlefield itself became a source of debilitating illness brought on by oppressive heat and dehydration. As spring faded into summer, though, it became clear that the sun-related illnesses were not the only threats from the physical environment. Scorching July heat was only one small part of the enigmatic swirl of atmospheric and climatic conditions that plagued field operations throughout the war. Unlike the sun, prolonged rain, drought, severe cold, and other elements might not be a direct cause of illness, but chaotic weather—perhaps the most unpredictable and potentially destructive force in the natural world—was about to play a major role in determining the course of the Civil War. From California to Virginia, soldiers and civilians alike would again be reminded that even the most carefully considered human efforts must accommodate the vagaries of nature.

Two

WEATHER

WINTER 1861–FALL 1862

Christmas 1861 brought gray skies and heavy rain to San Francisco. Precipitation had been falling in fits and starts since early November, often turning city streets into muddy troughs. For most of California's 380,000 residents, the stormy holiday weather probably seemed a fitting end to a gloomy political year. Secession and the South's victory at Bull Run had spawned a near rebellion among Confederate sympathizers in the southern part of the state. To the east, southern forces had invaded the New Mexico Territory and proclaimed the region a Confederate "Territory of Arizona." As if that were not troubling enough, various groups of Apache Indians launched strategic raids on both Union and Confederate armies in the contested region. Early in 1862, advance units of California volunteers began a treacherous journey across the Colorado Desert toward Fort Yuma on the Arizona border. From there, they hoped to beat back Confederate advances in the New Mexico Territory and protect Federal mail and transportation routes across the Southwest.[1]

As California prepared for war, the storms that began in late December ushered in a flood of near biblical proportions. For forty-five days that winter, rain rolled off the Pacific Ocean in wind-driven sheets, pounding already soggy terrain. Observers in San Francisco recorded thirty-seven inches of rain. A staggering 102 inches of precipitation inundated the Sierra Mountain village of Sonora. Sixty-six inches fell in Los Angeles (four times the *annual* average), turning the Santa Ana River into a four-mile-wide lake that covered a large portion of what had once been the Mojave Desert. Severe flooding in Sacramento forced the governor to travel to his inauguration in a rowboat before the state legislature abandoned the city and reconvened in San Francisco. In California's Central Valley, floodwaters submerged nearly

Downtown Sacramento during the great flood of 1862.
Courtesy California Historical Society.

3.5 million acres of the state's best farmland. Steamboats easily navigated the new 6,000-square-mile inland sea, delivering supplies to desperate farmers and ferrying the few surviving livestock to higher ground. William Brewer, a state surveyor and eyewitness to the storms, marveled at the destruction: "America has never before seen such desolation by flood as this has been."[2]

Analysts estimated the property damage in California at $10 million, an inconceivable number at the time, yet one that might be low. The more than 200,000 head of cattle that drowned or froze to death would have accounted for nearly half that total. The storms also ruined the grain and potato crops, washed away grapevines and orchards, and uprooted fruit trees. The raging rivers in the Central Valley picked up loose sand and topsoil and deposited the dirt downstream, where it formed a thick layer of sediment on formerly arable land. It would take months for any vegetation to return, raising the prospect of food shortages. In addition, rising waters destroyed thousands of homes, barns, mills, and granaries. Over the long term, the cost of recovery must have been many times higher than the initial estimate.[3]

As far south as the Arizona border, floods washed away or effectively isolated numerous army forts and outposts. Even when the waters retreated, commanders found it nearly impossible to locate sufficient forage for horses and mules, or food for their soldiers. Indians trying to stave off their own starvation stepped up attacks on white settlers and their few remaining livestock. The first Union volunteers heading for Fort Yuma had prepared for desert heat and drought, traveling in small, widely spaced groups so as not to tax the few available watering holes. Instead, the men encountered a deluge. Unrelenting rains and muddy roads slowed the march to a snail's pace. The Colorado River raged outside its banks and, in effect, turned Fort Yuma into an island. Thousands of dollars' worth of crucial supplies simply disappeared in the rising waters. Crippled by the storms, the "California Column" came to a quick halt, potentially leaving New Mexico to the Confederates—at least for the moment. The California troops would not move in earnest until April 1862.[4]

The origins of such turbulent weather can be traced to North America's dim geologic past. The process popularly known as continental drift positioned Civil War battlefields on a large landmass in the Northern Hemisphere, one that stretches to within ten degrees latitude of both the equator and the North Pole. As a result, North America has what scientists call a "continental" climate. In layman's terms, that means a climate marked by contrasts—warmer temperatures in its southern regions, cooler in its northern reaches. Such climates are generally dry, with most precipitation occurring in summer.[5]

However, as California soldiers and civilians could attest in 1862, that description can be more than a trifle misleading. Like everything else in nature, climates undergo constant change, shifting subtly toward colder or warmer temperatures and toward more or less precipitation. Known as "climate oscillations," these variations often have a profound effect on immediate weather conditions in North America. The oscillations most familiar to modern Americans are El Niño, a tendency toward warmer Pacific Ocean temperatures, and its counterpart, La Niña, which brings cooler currents to the region. Both are part of a larger pattern, known as the El Niño Southern Oscillation, or ENSO. It results from shifts in temperature and atmospheric pressure across a vast expanse of the Pacific Ocean, changes that can dramatically influence weather across the western and southern United States.

El Niño generally brings cooler, wetter weather; La Niña usually spawns warmer, drier conditions.[6]

Strange as it might seem, though, the historic floods that devastated the West Coast in the winter of 1861–62 actually occurred during a prolonged La Niña phase of ENSO. Between 1856 and 1865, the West and South endured ten years of substantially diminished rainfall. Conditions across the two regions proved so abnormally dry that climatologists (working with computer models and data gleaned from tree-ring analysis) routinely refer to the period as the "Civil War Drought." Thus, California got its worst rain in memory during one of the driest decades on record.[7]

No one knows for certain what happened in 1862 to trigger the deluge. But when ENSO moves into the La Niña pattern, it can spawn a much shorter oscillation known as the Madden-Julian Oscillation, or MJO. Certain phases of MJO can create conditions that favor the formation of a meteorological phenomenon known as an atmospheric river, a long narrow band of water vapor several thousand feet above the Pacific Ocean. Atmospheric rivers trap moisture from the tropics and funnel it toward drier regions. When the vapor hits high mountain ranges like the Sierras, the water condenses and falls as rain or snow. Most atmospheric rivers that affect California's weather are relatively small, but flooding can be severe. Today, Californians call such an event a "pineapple express," so named because of its ability to move water rapidly from Hawaii across the Pacific.[8]

Occasionally, though, shifting ocean temperatures and winds create trails of trapped water vapor hundreds of miles wide that extend for thousands of miles across the Pacific. Meteorologists call such a phenomenon an ARkStorm—"AR" for atmospheric river, "k" for 1,000, as in a 1,000-year event. The acronym also cleverly alludes to the flood that made Noah famous. An ARkStorm is precisely what happened in 1861–62. Fed by the inexhaustible supply of moisture from the tropics, a huge atmospheric river containing perhaps "as much water as ten to fifteen Mississippi Rivers," emptied its contents on the West Coast. No one in California had seen anything like it. Nor has anyone since, though some meteorologists theorize that the interval between ARkStorms might be closer to 200 years than 1,000. As California's stunned residents watched fertile agricultural valleys become giant lakes and deserts turn into muddy bogs, civilians struggled to stave off starvation and military commanders discovered that battle plans based on climatic norms proved useless.[9]

While the California Column waited for the sun, their Confederate adversaries in New Mexico yearned for rain. Effectively blocked by the Sierra Nevada, the ARkStorm of 1861–62 never made it to the interior of the New Mexico and Arizona Territories. As the air from the tropics rose over the Sierras, it cooled and condensed, dropping the bulk of its moisture on the western or windward side of the range. The air that advanced across the peaks to the eastern or leeward side proved much drier, leaving deserts there virtually untouched by precipitation. Known as the "rain shadow effect," the phenomenon is as old as the mountains themselves. It is primarily responsible for the arid climate of the Southwest and for the drier conditions on the Great Plains where the Rockies block moisture from the Pacific.[10]

Buoyed by Lt. Col. John Baylor's capture of Fort Fillmore in the heat of the previous summer and undeterred by the arid climate, the Confederate government planned to expand its control over New Mexico and Arizona. From there, southern leaders hoped to move on California, utilizing the supposedly thousands of Confederate sympathizers in the state. They sought to capture the gold and silver mines of the Sierras and eventually raise the Stars and Bars over several Pacific ports, including San Francisco. Leading this ambitious operation was Brig. Gen. Henry Hopkins Sibley, an old army officer with much experience serving in New Mexico.[11]

Sibley's army left San Antonio in October 1861 and crossed the Texas–New Mexico border in February 1862, moving beyond Fort Fillmore and defeating Union forces at the Battle of Valverde. He then pushed north along the Rio Grande deep into New Mexico, capturing Santa Fe and Albuquerque. By March, he made it to Glorieta Pass, a strategic site from which he might control access to one of the most important overland routes into the West: the Santa Fe Trail. For three days in late March Sibley's men engaged a Union contingent composed of tough, rowdy miners and adventurers from Colorado who had marched through freezing weather and blustery winds across the high plains to New Mexico. Though slightly outnumbered, the Confederates eventually took control of the pass. However, near the end of the third day's fight, a small Union flanking force stumbled upon the Confederate supply train, surprised the wagon guards, and destroyed all eighty wagons. It would turn the tide of the campaign.[12]

Unable to replenish their supplies in the desert, the Confederates could not press their advantage and quickly fell back, first to Santa Fe and eventually all the way to San Antonio. Long before the Union's California Column arrived in June, the Confederates abandoned their western campaign. Glorieta Pass proved to be the farthest western point of Confederate advance. Pointing to the mistakes that led to the loss of Sibley's supplies and the Rebel defeat, some historians have with considerable exaggeration called the engagement the "Gettysburg of the West."[13]

Overall, however, the Confederate failure to control New Mexico and Arizona owed as much to local weather, and its effect on food and water supplies, as to weapons and strategy. Sibley's army had set off from San Antonio in stages, one regiment at a time, in an attempt to make the most of the region's scarce grass and water. But before they reached New Mexico, the regiments had to divide into even smaller units. Moreover, Confederate forces had to advance (and eventually retreat) through the valley of the Rio Grande, the only reliable source of potable water. Forage for horses was almost nonexistent. According to one soldier, they rode through a west Texas landscape "where mesquite could not grow, cactus were drying up, and grass and such good things were not to be thought of."[14]

The Battle of Valverde in February centered on the main water source in that region. One Confederate soldier recalled that the Union defensive plan was "to hold the river and starve us out for water." The night before the battle, Confederates and their animals suffered overwhelming thirst. "The cry of the horses and mules in their agony was almost pitiful," recalled one Texan. Many mules broke away during the night searching for water and were captured by Union cavalry. Some soldiers were so desperate for water that they crawled through Union lines to fill their canteens at the river.[15]

As the Confederates advanced, they also went hungry, despite plenty of wild game that they could see but never get close enough to kill. After Sibley's men struggled up 5,300 feet of elevation to reach Albuquerque, they found that Federal forces had burned all their storehouses as they left the city. Almost no food remained. A sudden cold snap brought snow and sleet to that mile-high village that added to the brigade's woes. According to one soldier, they camped near Albuquerque for a week, during which they "buried fifty men." Had they remained longer, he explained, "we would have buried the whole brigade." Amid such miserable and highly changeable desert weather, and short on supplies after the destruction of his wagons at Glorieta Pass, Sibley decided he could not sustain the campaign and ordered a return to Texas. The march back east proved equally unpleasant. On the return to San Antonio, Comanche Indians—who resented all white presence in the region, regardless of uniform color—poisoned the available water sources. After one twenty-eight-mile march, Sibley's thirsty forces arrived at their destination to find that "Indians had filled the spring with dead oxen." They continued farther to a place known to have a good well, only to

Gen. Henry Sibley advanced along the Rio Grande into New Mexico during his 1862 campaign. He retreated from Albuquerque along the same route after losing his supply wagons at the Battle of Glorieta Pass. From Johnson and Buel, *Battles and Leaders of the Civil War*, 2:105.

discover that Comanches had "filled the well with dead wolves." In all they trudged fifty miles in the July heat without a drop of water.[16]

Men and animals collapsed along the route. Nearly all of them would have perished if a sudden thunderstorm had not come to their rescue that night. For two hours, wrote one parched soldier, "good, pure, soft, sparkling water poured down upon us." They had set out with 3,000 men and returned with fewer than half that number despite few battle casualties. Exposure to the elements, lack of food, and especially the dearth of water claimed the rest. Meanwhile, several hundred miles west, on the other side of the mountains, Californians still labored to repair damages from the worst flood in state history. Such were the vagaries of weather and climate on the North American continent in the winter of 1862.[17]

East of the Mississippi River, few soldiers and civilians took more than a passing notice of Sibley's plight or the Great California Flood. Easterners had their own problems with weather and war. The troubles began in Tennessee. Just outside Knoxville in the foothills of the Appalachian Mountains, the French Broad and Holston Rivers merge to form the mighty Tennessee. From there, the great waterway loses more than 500 feet in elevation as it meanders 650 miles through Chattanooga into northern Alabama, before veering back into Tennessee and flowing north into the Ohio River at Paducah, Kentucky. Those who lived along the Tennessee had often seen the river submerge the surrounding terrain. Four times during the 1840s major floods ravaged the Tennessee Valley; the worst occurred in 1847 and eclipsed all previous records for high water. However, with La Niña conditions in the Pacific, the river had been strangely placid since the early 1850s, and memories of the last flood seem to have been lost amid the political turmoil of the intervening years.[18]

The Tennessee is not the only stream born in the Appalachians that empties into the Ohio near Paducah. The Cumberland River, slightly smaller and equally prone to flooding, flows out of eastern Kentucky southwest through Nashville. From there it runs back into the Bluegrass State, where it parallels the Tennessee en route to the Ohio. Together the Tennessee and Cumberland provided relatively easy access to the interior of the Confederacy. With the onset of hostilities in 1861, both sides recognized the area's strategic importance. By early 1862, Confederate engineers had constructed two forts near the Kentucky border to fend off a Union invasion: Fort Henry on the Tennessee and Fort Donelson on the Cumberland. Though only twelve miles apart, the two southern bastions occupied strikingly different locations on their respective waterways. While Fort Donelson stood on a high bluff 100 feet over the Cumberland, Confederate engineers located Fort Henry squarely in a floodplain, a mere 20 feet above the banks of the Tennessee. Even so, Gen. Albert Sidney Johnston, commander of the Confederacy's western theater, characterized Fort Henry as "a strong work," well situated to defend a crucially important water route into the South.[19]

Had the weather patterns of the previous decade held a bit longer, Johnston's faith in Fort Henry might have been justified. However, as the residents of California had already discovered, 1862 was no ordinary weather year. Beginning in January, heavy rain on the western slopes of the Appalachians and throughout the length of the Tennessee Valley sent the river surging out of its banks. In early February, with the Tennessee approaching flood stage, Gen. Ulysses S. Grant launched a combined army and navy attack intended to drive the Confederates out of their defensive positions. Muddy roads slowed the infantry's progress and delayed its attack, but that made little difference at Fort Henry. By the time Union gunboats arrived at the fort, the Tennessee River was some thirty feet above its low-water mark—about ten feet too high for the ill-sited fortress to survive. Only nine of Fort Henry's seventeen guns remained dry and functional. Rising waters soon swamped the powder magazine, leaving the Confederates without adequate ammunition. Meanwhile, the surprised Yankee gunboat commanders, high and dry on their vessels, found themselves eye-to-eye with southern soldiers in the fort, who were trying in vain to fend off both the enemy and the rising river. Fort Henry's commander, Confederate brigadier general Lloyd Tilghman, realized the hopelessness of the situation. After ordering most of the garrison to march overland to Fort Donelson, he held out for seventy-five more minutes (during which time he helped service the fort's few functional guns himself) before he surrendered. Tilghman later complained that the fort's location was so miserable that the "history of military engineering records no parallel to this case."[20]

As Grant's forces turned their attention toward Fort Donelson on February 12, unusually mild temperatures led many exuberant but inexperienced Union soldiers to discard their heavy overcoats and blankets. The next night, as the northern troops camped outside the fort, the temperature dropped to 12 degrees Fahrenheit and a cold rain turned first to sleet and then to wind-driven, face-stinging snow. Worried about giving away their location, officers

Union gunboats bombard Fort Henry on February 6, 1862. The Tennessee River was so high that the defenders of the flooded fort were at eye level with the Union vessels. Courtesy of the Naval History and Heritage Command, Washington, D.C.

refused to allow freezing Union men to build fires to brew coffee or provide a buffer against the cold. Confederates inside the fort fared little better. One defender recalled it being so cold "that we could barely stand still during the hour of midnight." Extreme cold discouraged much fighting on February 14, but the two forces clashed the next day before the snow and piercing winds resumed.[21]

During the frigid night of February 15, Gen. John Floyd, the commander at Fort Donelson, decided it was useless either to hold out longer or to break out for Nashville. Given the weather, one of the fort's ranking officers estimated that nearly three-quarters of the garrison might perish from frostbite or pneumonia in an attempted retreat. Nevertheless, about 5,000 troops managed to flee. More than half of them, cavalry troops under the command of Gen. Nathan Bedford Forrest, crossed the icy tributaries of the Cumberland and escaped. One of those who fled was Floyd, who had turned over command to Gen. Simon Bolivar Buckner. Buckner surrendered the remaining 15,000 Confederate soldiers to Grant on February 16, 1862.[22]

As the Confederate defenses collapsed, the Union navy immediately exploited the opening of the Tennessee River. Gunboats steamed all the way to Florence, Alabama—200 river miles away—destroying ships and supplies all along the way. The capture of the forts opened important invasion routes into the western reaches of the Confederacy and deprived the South of nearly one-quarter of the men assigned to protect that vital territory. Southern units from locations throughout Tennessee and Kentucky hastily fell back all the way to Corinth in northern Mississippi.[23]

The Union army of the Ohio, under Gen. Don Carlos Buell, captured Nashville, a vitally important southern supply depot, while the Army of the Tennessee moved along its namesake river to Pittsburg Landing, just twenty miles north of Corinth. Union commanders chose the site largely because it perched on high ground safely above the river, which, in flood again, had risen fifteen inches in a twenty-four-hour period in early March. After some misunderstandings with Union western theater commander Gen. Henry Halleck that led to his being temporarily relieved of command, a reinstated Grant finally took charge of the Army of the Tennessee at Pittsburg Landing. There, Grant waited with approximately 40,000 troops for the anticipated arrival of Buell's army of 30,000. Once these forces joined, Halleck planned to take direct command to attack the entrenching Confederates at Corinth. Grant was confident that a decisive victory in this one last "big fight" would end the war, if it would just stop raining.[24]

Trying to shift the South's fortunes, Johnston decided to launch his own offensive before Union forces could unite. He set the attack for the morning of Friday, April 4. However, heavy rains and confusing marching orders slowed the southern army to a crawl. As a result, the attack began two days late, on April 6—a critical delay that allowed Union reinforcements to join Grant. In the ensuing Battle of Shiloh, Confederates enjoyed initial success, though Johnston suffered a mortal wound in the fighting on the first day. But Buell's troops had arrived by that evening. On the night of April 6, they ferried across the river—in a blinding rainstorm, no less—and helped drive the Confederate army away on April 7, securing a Union victory in a battle that foreshadowed the grim fighting to come in the war, as the two foes suffered more than 23,000 casualties combined.[25]

The Confederates retreated to Corinth along the same atrocious roads by which they had approached Shiloh. One Louisiana soldier remembered the retreat to Corinth as an exquisite type of water torture. The men marched "amidst a terrible storm of rain and sleet." When yet another torrent opened on the night of April 7, the soldier wrote, "I was too weary to mind that. Flinging myself down in the mud, and rain, I slept the dreamless sleep that only comes to soldiers when they are dead tired." Corinth itself was "a very disagreeable place," wrote Alabama soldier Joshua Callaway on May 5. "It is shoe mouth deep in the nastiest mud I ever saw, & the water is mean enough to kill an alligator." The bad water led to rampant sickness, and soldiers died with frightening rapidity. A shaken Callaway recalled three months later, "I shudder when I think of the suffering and death that I witnessed while at Corinth. . . . my ears were constantly saluted by the groans of the sick & dying."[26]

The weather after the Battle of Shiloh did not aid Union troops either. The rains not only made the roads impassable for supplies but also washed out the shallow graves dug to bury those killed in the battle. The stench of decaying human and animal corpses nauseated the soldiers. The constant washouts made it nearly impossible to find clean water, giving rise to dysentery and prompting a constant search for higher ground. Under General Halleck, the combined armies eventually moved slowly toward Corinth and captured the city on May 30, 1862, after the weakened Confederate army had evacuated it.[27]

Like the California floods, the wet weather that facilitated Grant's success at Fort Henry and Fort Donelson, and led to the bloodbath at Shiloh, occurred in the middle of the "Civil War Drought." However, just as in the West, everyday conditions often varied considerably from the climatic trend. During a La Niña winter, the polar jet stream—the fast-flowing swath of air that transports weather fronts west to east across North America—typically dips into the Ohio Valley. When that happens, extended periods of heavy rain or snow can fall sporadically across Kentucky and Tennessee even in years of less-than-average annual precipitation. The Appalachian Mountains also intercept systems moving in from the West—just as they did in 1862—and deposit the moisture as rain or snow on the western slopes. The eastern ridges often receive less precipitation, a micro version of the process that creates drier climes east of the Sierras and Rockies.[28]

ENSO oscillations were not the only long-term trends affecting the wild weather of 1862. Atmospheric conditions across the country might have been more volatile due to another, longer-term climatic shift known as the Little Ice Age. Between 1300 and 1870, winters in the Northern Hemisphere—especially Europe and North America—were much colder on average than in the twentieth century. Theories about the origins of the Little Ice Age run the gamut from unusual sunspot activity, to atmospheric fluctuations in the North Atlantic, to volcanic activity and decreases in the human population. Whatever its cause, as the Little Ice Age drew to a close in the 1860s, the overall trend toward warmer temperatures likely increased the potential for stormy weather across the continent, even as La Niña brought on a drought.[29]

For Americans during the Civil War, daily weather excited much interest. Soldiers wrote often of local conditions, but their descriptions were mostly general and emphasized the unusual (excessive heat, cold, rain, snow, wind, and violent storms). However, thanks to a Presbyterian minister (he preferred the designation O.S.P. for "Old School Presbyterian") living in Washington, D.C., we have at least one systematic record of temperatures and rainfall. Keenly interested in weather, the Reverend C. B. Mackee carefully noted the temperature at 7:00 a.m., 2:00 p.m., and 9:00 p.m. almost daily between October 1860 and June 1865. Compiled and supplemented with other data by historian Robert K. Krick, Mackee's observations reflect the complex relationship between climate and weather, including widely variable local conditions. The lowest January temperature Mackee recorded, six degrees Fahrenheit, occurred twice, at 7:00 a.m. in 1861 and 1864. In the latter year, the reading came just one day after a reading of forty-five degrees at the same time, a thirty-one degree-drop in just twenty-four hours. In January 1862, the 7:00 a.m. temperature fell from fifty degrees to twenty-four degrees in two days. Likewise, temperatures recorded at 2:00 in July varied widely, from a spring-like sixty-six, recorded in 1861, to a blistering high of ninety-eight in 1864. Accounts from other sources about wind, rain, snow, and sleet are less systematic but suggest a similarly chaotic pattern, with a slight trend toward more precipitation in the spring and summer.[30]

Such variations point to a disquieting truth: local weather — during the 1860s or the present — is a phenomenon that can be observed and explained but is stubbornly difficult to anticipate. In general, conditions like those recorded by Mackee in the 1860s result in part from climatic oscillations like ENSO and the Little Ice Age. But the daily weather in Washington, D.C. (or anywhere else in Civil War America), stemmed from myriad other influences, including local variations in topography, barometric pressure, and temperature. In 1960, meteorologist Edward Lorenz famously termed this phenomenon the "butterfly effect," suggesting that something as innocuous as a butterfly flapping its wings in Asia could profoundly influence weather patterns in New York City. Lorenz was an early proponent of what today is known as chaos theory. Simply stated, chaos theory holds that in a complicated system like Earth's atmosphere, even a slight change in an initial action can drastically alter final results.[31]

Acknowledging the importance of weather to Civil War history inevitably introduces that element of chaos to the battlefield. Wherever soldiers ventured, whatever tactics and weapons they employed, the success of their operations often hinged on chancy atmospheric conditions that fluctuated wildly and without warning. By the spring of 1862, chaotic wartime weather had profoundly affected military operations from California to New Mexico to Tennessee—and the year had scarcely begun.

As winter turned into spring along the East Coast, President Abraham Lincoln and Gen. George B. McClellan, commander of the Union Army of the Potomac, made plans to attack the Confederate capital at Richmond, Virginia. Situated on the James River where the Virginia Piedmont uplands meet the more level coastal plain, elevations in Richmond range from near sea level along the shores of the James to more than 300 feet in the city's western reaches. Confederates defending the city thus enjoyed the advantage of occupying elevated terrain. Only a hundred miles or so from Washington, Richmond was the political and manufacturing center of the Confederacy. Given its strategic importance, both Lincoln and McClellan believed that taking the city might bring a quick end to the war.

However, the two men differed about the best way to attack Richmond. The president preferred to approach Richmond from the north by having McClellan move against Confederate general Joseph E. Johnston's army, which still held the region around Manassas after its victory there the previous summer. From Manassas, the Union army would then continue south overland to attack Richmond. Lincoln believed such a direct strategy had a better chance of success and made it easier to defend Washington in the event of a Confederate counteroffensive. McClellan had other ideas. Beginning in February 1862, he devised an elaborate plan to transport his troops by boat down the Potomac to the Chesapeake Bay, then go up the Rappahannock to Urbanna, Virginia, about fifty miles northeast of Richmond. From there he could outflank Confederate defenses. While the frustrated president debated with his commander, General Johnston unexpectedly pulled his troops back from Manassas to Fredericksburg on the Rappahannock. The move negated McClellan's Urbanna plan, because Johnston would be close enough to respond rapidly to the Union's army's movement.[32]

Refusing to concede any strategic point to Lincoln, who he felt had only an amateur's understanding of military operations, Little Mac (as McClellan's men called him) quickly decided to move his army to Fort Monroe at the eastern tip of the Virginia Peninsula, a move that even he had earlier described as a last resort, "the worst coming to the worst." Supported by gunboats on the James and York Rivers, which flowed along the southern and northern edges of the long and relatively narrow stretch of land, McClellan intended to lead his formidable army west, straight up the peninsula, destroying Confederate defenses along the way. He planned to cross the Chickahominy River, a major tributary to the James, and lay siege to Richmond. Long a serious student of siege warfare, he believed that a swift advance followed by heavy bombardment could bring Richmond to its knees. Lincoln remained less sanguine about the plan, but with his notoriously slow general finally on the move, the president reluctantly agreed to the ambitious strategy.[33]

For a former engineer who prided himself on his planning abilities, McClellan apparently gave little thought to the peninsula's geography and climate as he contemplated the campaign. His chief engineer, Brig. Gen. John G. Barnard, remarked that the Virginia Peninsula was "a terra incognita" for the army planners. "We knew the York River and the James River and we had heard of the Chickahominy," Barnard admitted, "[but] this was about the extent of our knowledge." McClellan's unfamiliarity with the terrain would reveal itself early and often. Less than a week into the campaign he confessed, "The topography of the country was very different from what had been supposed." The farther his army advanced, the more the soldiers' ignorance showed. In May, McClellan admitted that his progress was so slow because "we have to feel our way everywhere; the maps are worthless."[34]

The unfamiliar landscape had been shaped as much by people as by nature. The region's warm summers and plentiful spring rains had once proved ideal for tobacco cultivation during the colonial period. By the 1860s, though, peninsula soils had been depleted to the point that farmers had turned from tobacco to wheat, corn, oats, and livestock—all of which could thrive on land that had been cultivated for more than two centuries. Over time, the landscape had evolved into a patchwork of cleared farmland, brushy old fields, and thick second-growth forests of loblolly pine, sweet gum, and red and white oak. Within those deep woods, the peninsula's plentiful springs and creeks occasionally formed nearly impenetrable swamps. To get their crops to market, local farmers relied on a small but viable network of sandy roads. As McClellan planned his attack on Richmond, he knew that he would have to depend on those same routes to move men and matériel. Confident in reports of the region's "good natural roads" that would support his wagons and half-ton artillery pieces, he believed that he could get his troops into position to turn their guns on Richmond.[35]

Southern observers more familiar with the region thought otherwise. In early February 1862, a month and a half before the first of McClellan's men left for the peninsula, the editor of the *Richmond Examiner* explained that, "Unlike the winter of the North, which is a season of dry snow, firm roads and facile locomotion, the winter of the South is a season of rains, mud, and mire." He then speculated, "If his [McClellan's] troops push their invasion into the interior, they will have to become amphibious, and borrow some of the qualities of alligators and mud turtles. Instead of marching, they will have to wade against the secessionists." His words proved prophetic.[36]

Largely ignorant of the environment he was entering, by late March 1862, the Young Napoleon (another nickname that McClellan relished) had amassed over 110,000 men, 40,000 horses, 4,000 wagons, and nearly 250 pieces of artillery at Fort Monroe. The general himself arrived in April. Witnessing the massive army—an assemblage of people and animals larger than any city in Virginia—a British observer noted that McClellan came ashore with "the stride of a giant." Little Mac seemed keen to move quickly. The general disembarked 60,000 of his soldiers off their transports and immediately began to march toward Yorktown, about twenty-five miles away. His enthusiasm lasted exactly one day.[37]

Having progressed fewer than fifteen miles, McClellan awoke to a driving rain on April 5. As he contemplated the foul weather, the general was astonished by a message from his advance corps commander, Erasmus D. Keyes, explaining that Confederate troops had formed a strong defensive line along the Warwick River. A tributary to the James, the Warwick was a sluggish stream that began half a mile south of Yorktown and flowed south eight miles, bisecting the peninsula. McClellan expected to encounter the Warwick as he moved toward Richmond, but the general did not expect to find the stream blocking his path. Having studied a badly flawed map drawn up by topographical engineer Col. Thomas Jefferson Cram, McClellan thought the river flowed primarily west to east parallel to the James. Indeed, looking at the faulty map, McClellan had even considered deploying gunboats on the Warwick to protect his left flank. Only in the middle of a rainstorm on April 5 did he realize the river presented a difficult obstacle.[38]

The Confederates, under the command of Gen. John Bankhead Magruder—known as "Prince John" for his affinity for theatrics and his penchant for living above his means—had chosen their position carefully. The thick leafy woods along the Warwick provided ideal cover for the Rebel forces. Concealed in the forests, Magruder's 13,000 men could maintain an advantageous position perpendicular to McClellan's path and easily observe

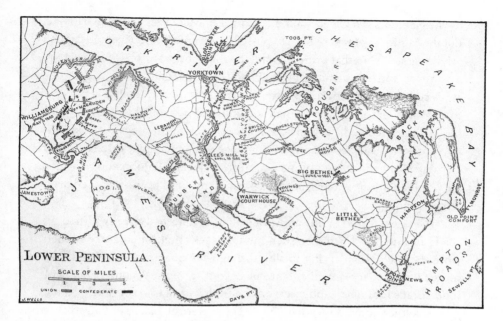

Gen. George McClellan disembarked his troops at Fort Monroe in April 1862 and advanced up the peninsula. He was stymied by a misunderstanding of the terrain and the strength of Confederate defenses under Gen. John Magruder. From Johnson and Buel, *Battles and Leaders of the Civil War*, 2:188.

Union troop movements. Gen. Lafayette McLaws, a division commander in Magruder's army, put it best: "The country is so much cut up by these arms of the sea, called rivers, with their accompanying marshes and boggy freshwater tributaries that it is impossible to move through it except along the main roads." On the peninsula, McLaws believed, "the tremendous odds against us can not be so formidable as elsewhere in more open country."[39]

Prince John also manipulated nature to his advantage. For years, local residents had maintained a series of dams on the Warwick to prevent flooding and power their grist and saw mills. In a savvy strategic move that took advantage of the local geography, Magruder built three more dams that caused the overflowing river to flood established fords, submerge nearby roads, and create small swamps in the surrounding forests. If McClellan continued his advance, he would have to make use of the remaining few crossing points, all carefully guarded by Magruder's men ensconced in defensive works and supported by artillery. As Keyes warned McClellan, to attack the Confederates in that position, would result in "an enormous waste of life."[40]

Betrayed by a faulty map, surprised by the flooded terrain, and hounded

Weather

55

by the elements, McClellan still outnumbered Magruder nearly five to one. But the cautious general refused to push that advantage. He accepted Keyes's original assessment that a direct assault on the Rebel position would cost too many lives, and he convinced himself that Magruder had a much larger force at his command. (Prince John often marched his men in repeating patterns to convince Union observers that he had far more troops at his disposal.) McClellan opted to dig in and lay siege to Yorktown, a strategy that required him to move his heavy guns to within range of Confederate defenses. After learning of the movement of the Union army, Joseph E. Johnston relocated his Confederate army to Magruder's position by April 14 and assumed overall command.[41]

The skies cleared briefly on April 6, but the rains had already turned the roads around Yorktown into such muddy tracks that the infantry soldiers wore themselves out trying to slog through the quagmire. It proved even more difficult to move artillery into place for the siege. Wagons loaded with food, forage for horses, and reserve ammunition sank into the muck. As one Union soldier noted on April 6, "The wagons were all halted or had stuck in the mud. . . . The mules wallowed in the mud and cleared themselves completely from their harness in one or two rolls. There was vehement swearing and lashing of whips by the teamsters when we would go 500 feet more ahead and stick again. This traveling was very tedious and did not help to keep any in good humor."[42]

Tedious indeed. It took Union forces a month to prepare for the siege, and according to observers up and down the peninsula, rain fell on sixteen of those thirty days. From April 7 to 10, it rained for four days straight before the storm cleared out, leaving snow flurries and freezing temperatures in its wake. By the time McClellan finally got his siege guns into place, the Confederates had abandoned their waterlogged defenses. Johnston never held a high opinion of the position. Unimpressed by the efforts of Magruder's engineering officers in constructing the defensive works, Johnston snidely reported to Lee, "No one but McClellan could have hesitated to attack." After a brief delaying action at Williamsburg on May 5, 1862, during which it rained steadily for almost another twenty-four hours, the Confederate army began a retreat toward Richmond. Union forces slowly followed—and watched the skies.[43]

Though the sun returned briefly as McClellan's forces left Williamsburg, the near-month of early spring rain left local roads muddy and treacherous. Getting Union artillery through the mud proved especially difficult. McClellan brought forty-four artillery batteries to the peninsula. Each bat-

The Army of the Potomac, slowed by rain and mired in mud, between Williamsburg and Richmond, May 1862. From *Harper's Weekly*, June 7, 1862.

tery typically consisted of six cannons, most weighing 1,200 pounds each. Moving each of the guns required a six-horse team. In addition, each battery needed another eighty horses to pull the supporting ammunition caissons and supply wagons. Roads that had been adequate for transporting local crops to market now gave way under the massive weight. Every hoof and wagon wheel cut deeper pockets and ruts that made the roads worse. The tiring conditions of the muddy roads, coupled with the lack of adequate forage for the horses, taxed the ability of even the strongest teams. On May 5, McClellan wrote Secretary of War Edwin Stanton, noting in near disbelief that "several of our batteries are actually stuck fast in the mud. The men have done all that could be done."[44]

From May 14 through May 28, observers in various locales across the peninsula recorded thirteen days of rain. "The roads are beyond description," wrote a Union chaplain. "Just imagine the worst roads possible, and then believe [these] are ninety-nine times worse than your imagination and you may then come near the truth." One Confederate soldier concurred that with the "roads awfully muddy," the men in his unit were "nearly all broken down with fatigue, hunger, & want of sleep." A South Carolina soldier wrote on May 13, "I never saw men suffer so much in all my life. They were half fed & marched almost to death."[45]

Union and Confederate soldiers often collapsed from exhaustion trying to slog through the mud. Estimates suggest that a 143-pound man (the average weight for a Civil War soldier, according to one study) marching in mud burns approximately 545 calories per hour. Marches of eight to ten miles could take as many hours in the dreadful weather conditions.[46] Burning at least 5,000 calories a day, the advancing Union soldiers often had limited access to provisions since most of the supply wagons remained stuck in the mud to the rear. McClellan felt for his troops, confessing to his wife on May 6, "It is with utmost difficulty that I can feed the men, many of whom have had nothing to eat for 24 hours & more." Some of the equally weary retreating southern soldiers also went hungry because their commissary wagons sank into the mud, "many being compelled to throw out rations in order to get along." Occasionally McClellan's troops came across abandoned Confederate wagons with dead mules "lying on their backs, half smothered in mud, with their feet sticking out of it."[47]

Though neither the Young Napoleon nor his Confederate adversaries knew it, the deep mud stemmed from geologic events that occurred tens of thousands of years earlier. McClellan's route to Yorktown and Williamsburg took him straight across what geologists call the Norfolk Formation, a sedimentary structure laid down in the Pleistocene epoch (between 2.5 million and 11,700 years ago). Deposited by ancient rivers and estuaries, the sediment is best described as a kind of clayey sand. Because of the clay content, it drains poorly and is prone to flooding, especially in relatively flat terrain like that of the Virginia Peninsula. In places, McClellan's route also crossed the Windsor Formation, an older geologic unit that, where it intersected the general's path, has properties similar to the Norfolk Formation.[48]

In the intervening millennia between the formation of the sedimentary layers and McClellan's campaign, regular flooding of the York and other nearby rivers laid down alluvium near the surface that also contained a mix of sand, clay, and organic material. Under moderately rainy conditions, the roads, which to the eye appeared sandy, could easily support farm wagons and light loads. However, during prolonged heavy rains, like those that fell on the peninsula in the spring of 1862, the clays of the soil and sediment beneath tend to absorb water. If agitated or compressed—in this case by thousands of horses, heavy artillery, and supply wagons—the clays undergo a process technically known as liquefaction. In layman's terms, the clay becomes an incredibly sticky deep mud that behaves more like a liquid than a solid, the sort of miry mess that could swallow wagons and suffocate mules. One geologically minded Union artillerist aptly described the soil as one or two feet of sand on top of marl on top of clay. "The immense rains we have had all this spring, sinking directly through the sand and finding no outlet from the marl, have converted it into the consistency of soft mortar," he wrote. "When a heavy substance once breaks through the top soil, there is nothing to stop its sinking until it reaches the hard clay."[49]

Unaware of the geologic history working against him, a frustrated McClellan simply referred to the roads and weather as "infamous," "frightful," or "execrable," pausing occasionally to note "another wet horrid day!" Still both armies slogged toward Richmond. As they drew closer to the capital, Union and Confederate soldiers found firmer ground (geologically speaking) beneath their feet. Near Richmond, the underlying sediment is coarser sand and gravel that drains better than the clayey sand to the east, allowing rainwater to pass through at a higher rate. The roads likely dried out quicker and were generally more passable than those farther down the peninsula. By May 24, Little Mac's forces had advanced as far as Mechanicsville, only five miles from Richmond — close enough, in fact, to hear church bells ringing in the city.[50]

Almost immediately, however, McClellan found himself facing another environmental obstacle. Perhaps because he expected reinforcements to join him from the north, McClellan divided his army as he approached Richmond, locating more than half his men north of the Chickahominy River and the remainder south of the waterway. A relatively narrow, sluggish stream that begins northwest of Richmond, the Chickahominy flows southeast for eighty-seven miles through a low flat valley before it empties into the James River. Level bottomlands along the Chickahominy make it especially prone to flooding even during periods of moderate rain. By the time McClellan arrived in late May 1862, the wet spring had already forced the Chickahominy out of its banks and turned the surrounding bottomland into a mile-wide swamp. Then, on the night of May 30, after two days of clear weather, some of the heaviest thunderstorms of the season rolled across Richmond and the surrounding area. The unexpected downpour and ensuing flood delayed construction on some bridges and washed out others across the Chickahominy, leaving each branch of McClellan's divided army to fend for itself.[51]

Sensing an opening that might allow Confederate forces to engage fewer than half of McClellan's troops, Johnston launched a surprise attack on Federal forces at Seven Pines, south of the Chickahominy on May 31. Suddenly on the defensive, McClellan rushed in troops from north of the river to stave

The Fifteenth New York built this military bridge across the Chickahominy River during McClellan's advance on Richmond in June 1862. Courtesy Library of Congress.

off Johnston's attack. It was, however, a dangerous maneuver, requiring the reinforcements to cross a precarious footbridge over the swollen stream only hours before the bridge disappeared in the raging current. More troops from the south side of the Chickahominy also made their way to the front, but they arrived without artillery, "it being impossible to move forward through the deep mud."[52]

This time the weather played no favorites. The rain was heavy enough to complicate travel even on the better-drained roads around Richmond and prevented the Confederates from getting some of their artillery batteries within easy range of McClellan's army. In addition, communication problems within the southern leadership led to a confused attack that, in the end, failed to drive Union forces away from Richmond. Even so, the two-day battle at Seven Pines cost the two sides over 11,000 combined casualties. Among that number was Johnston, who suffered a debilitating wound that forced him to give up command of the southern forces. From that point on, the South entrusted the defense of Richmond and the future of the Confederate army to Robert E. Lee.[53]

Had McClellan been able to move his heavy siege guns into range of Richmond after Seven Pines, Lee might well be remembered as the Confederate general who lost the capital and perhaps the war. Instead, it rained—again—almost incessantly for the first four days of June. After a clear day on June 5, wet weather returned on four of the next five days. Already seething because Lincoln had denied him all the reinforcements he desired, McClellan now railed against the elements. "It is again raining hard & has been for several hours! I feel almost discouraged," McClellan wrote to his wife on the night of June 10. "It is certain that there has not been for years & years such a season," he lamented. "I am quite checked by it." Lee faced the same conditions; he echoed McClellan when he wrote Jefferson Davis, "You have never seen roads like those in the Chickahominy Bottom." But he understood that the longer the rain stymied McClellan, the better the Confederate chances for a successful counterattack.[54]

Following Seven Pines, soggy conditions brought fighting to a near standstill for three weeks as McClellan deemed the ground too wet to move his heavy guns to the front. While the rains fell, Lee planned carefully for an offensive against the Union troops. He first sent his cavalry commander, James Ewell Brown ("Jeb") Stuart to scout McClellan's dispositions. Upon discovering that McClellan had weakened his right flank, Lee recalled Stonewall Jackson's troops from the Shenandoah Valley to spearhead an attack against McClellan's remaining forces north of the Chickahominy. On June 26, he launched a bold attack against McClellan's right flank north of the Chickahominy near Mechanicsville. Though the battle proved a tactical failure for Lee, it turned into a strategic success. Convinced that Lee's army outnumbered his (it did not), McClellan decided late that night to move his army south across the peninsula to Harrison's Landing on the James River. There, Union forces might be resupplied by water and protected by Union gunboats. In what would become known as the Seven Days' Battles, Lee repeatedly attacked McClellan's army as it moved to the James. During the retreat, McClellan's army fought four separate battles—at Gaines Mill on June 27 (a costly Confederate victory), Savage Station on June 29 and Glendale on June 30 (both tactical draws), and Malvern Hill on July 1 (a Union victory). McClellan suffered nearly 16,000 battle casualties that week, out of 114,000 troops, while Lee lost slightly more than 20,000 of the approximately 90,000 soldiers in his army.[55]

Violent thunderstorms pelted Union troops on the nights of June 29 and July 1 during the retreat, again turning the roads into rivers of mud. As Little Mac's men knew all too well, slogging across such ground could exhaust even the most intrepid soldier. Pvt. Robert Knox Sneden noted that the roads were better drained than those encountered further east on the peninsula, and he recognized their good fortune: "If we had to go over the red mud roads we would have hopelessly stuck in the mud all night and the enemy would have overtaken us." Still, it was a rough slog, as ten horses had to be harnessed to each artillery piece to get it through the mud. While burning thousands of calories during the retreat, most soldiers likely consumed fewer than 500 calories per day during that stretch. Much of their caloric intake came in for the form of coffee and hardtack, neither of which proved particularly nutritious. As the army's medical director wrote of hardtack: "This bread is difficult to masticate, is dry and insipid, absorbs all the secretions poured into the mouth and stomach, and leaves none for the digestion of other portions of the food." Worse, the coffee functioned as a diuretic and together with the dry baked cracker could bring on dehydration. After four days of marching and surviving on those rations, many of McClellan's men were on the verge of collapse. The retreat wore out animals, too. "Many horses and mules fell from exhaustion," Sneden noted. He added laconically, "The confusion which attended the march . . . was very considerable and demoralizing."[56]

Animals that survived the brutal pace added a peculiar hazard of their own. To be sure that Confederate forces did not capture the Union beef supply, nearly 2,500 head of cattle walked ahead of the fighting soldiers. In addition, thousands of horses pulled Union artillery and supply wagons. Those beasts chewed up the roads with their hooves and left thousands of noxious "pies" in their wake, directly in the path of the soldiers. The generally rainy weather meant that soldiers had to trudge shin-deep through a stinking stream of mud and excrement. Dehydrated and near exhaustion, Union soldiers were "glad to drink rainwater which had settled in the wheel ruts made by the passing artillery," and, of course, all the associated fecal bacteria that lived therein. Many soldiers ducked out of line to relieve themselves on the side of the road, adding still more contaminants to the available water supply. Small wonder that so many soldiers were sick by the time they reached their destination.[57]

The weather, poor diet, and numerous hardships fostered a variety of illnesses for soldiers in the region. Waterborne illnesses such as typhoid, diarrhea, and dysentery had begun plaguing the Union soldiers soon after they landed on the peninsula. In that region the water was particularly ill-suited to large numbers of men camping for an extended time, as Confederate troops had discovered the previous year. But many Union officers and surgeons noted a spike in sickness along the Chickahominy—the myriad disorders earning the sobriquet "Chickahominy Fever." Gen. Régis de Trobriand noted that the army's "sanitary condition became worse from day to day." Union surgeons sent sick soldiers to hospitals on the York River, where transports took the sickest of them off the peninsula. "But these transports," Trobriand recalled, "though making continual trips, were not able to take away the sick as fast as they arrived." Months later, many Union soldiers still battled chronic diarrhea, which, one Vermont soldier lamented, "we contracted in the swamps of the Chickahominy, and which saps the foundations of one's strength, and makes his existence a lingering duration of misery."[58]

The return of an old scourge only made matters worse. Both major types of malaria had been endemic on the peninsula for roughly 150 years. However, during the early and middle decades of the 1800s, the disease had been on the wane. The local population acquired immunity to common malarial parasites, and increased clearing and tillage restricted breeding sites for anopheline mosquitoes. That changed in 1862 as the converging armies provided tens of thousands of new hosts for parasites carried by the winged pests, and the rain created a perfect habitat for them. Even McClellan was not immune. He battled a recurrence of what he called his "Mexican disease," probably the malaria he contracted during the Mexican War. As medical and agricultural historian G. Terry Sharrer has explained, "The fighting was actually a secondary terror, happening in the midst of an unprecedented pandemic—not of one disease widely spread, but of many, all at once."[59]

Nutritional deficiencies compounded Union problems. Army medical personnel found it implausible that scurvy could appear on the peninsula, since the daily supply of food for soldiers included dried fruits and vegetables. However, surgeons had not counted on the dietary preferences of the individual soldiers. When six men in one brigade were diagnosed with scurvy on June 14, the army's chief surgeon called on the commissary director to see why he was not furnishing the men with the required vegetables. He discovered that "the men very generally refused to use the desiccated vegetables; the [commissary] had an abundance of them, and could not get rid of them." He ordered large supplies of lemons and reiterated calls for officers to make their men eat the hated dried vegetables. But Lee's attack on the Union army began before these efforts could be enforced.[60]

The new Union medical director, Jonathan Letterman, who joined the army on July 1 and accompanied it to Harrison's Landing, could scarcely believe the weakened state of McClellan's men. "The malaria from the borders of the Chickahominy and from the swamps throughout the Peninsula,"

Letterman wrote, "now began to manifest its baneful effects upon the health of the men." Marching nearly nonstop, the troops often struggled to prepare food. "They had little time for sleep, and even when the chance presented itself it was to lie in the rain and mud." Additionally, he declared that widespread illnesses such as scurvy "undermine the strength, depress the spirits, take away the energy, courage, and elasticity of those who do not report themselves sick, and who yet are not well. . . . In this way it had affected the fighting power of the army, and much more than was indicated by the numbers it had sent upon the reports of the sick." Medical officers also succumbed to illness, leaving fewer doctors to treat the increasing number of sick and wounded. At Harrison's Landing, sickness incapacitated more than 20 percent of McClellan's army. Disease affected the highest ranks as well. McClellan was annoyed that more than twenty generals applied for sick furloughs in early July. A month later, 10,000 soldiers remained incapable of taking the field.[61]

A higher percentage of soldiers in the Army of the Potomac were sick in July 1862 than in any other month of the war. Of the 106,069 soldiers listed on its rolls, more than 40,000 cases of illnesses and ailments of varying degrees were reported, with an astounding 19,776 soldiers afflicted with diarrhea or dysentery. Only October 1862 would find more soldiers suffering from diarrhea and dysentery (21,234); however, not only was that from a much larger army (171,258 soldiers) but many were chronic cases that had developed on the banks of the Chickahominy. On July 4, Letterman found McClellan's troops so sick and exhausted that he drew up a lengthy prescription for their recovery, including a diet rich in fresh vegetables, proper shelter from the elements, and plenty of rest. Unfortunately, Harrison's Landing—a flat plain surrounded by swamps and overrun by aggressive mosquitoes and flies—was not a particularly healthy place for the army to camp either, and sickness continued through the summer. Though McClellan called for reinforcements to launch another attack on Richmond, some of his generals strongly disagreed. Erasmus Keyes wrote to Lincoln strenuously urging the president to withdraw the army from the unhealthy environs. Keyes implored, "To bring troops raised at the North to the country in the months of July, August, and September would be to cast our resources into the sea." McClellan finally received orders to abandon the peninsula on August 3, 1862, ending the campaign.[62]

McClellan had his chances to turn things around. At Malvern Hill, the bloody finale to the Seven Days, Lee's forces suffered 5,000 casualties; McClellan's losses totaled less than a third of that number. Many Union officers strongly urged their commander to launch a counterattack against the weakened Confederate army after this victory. When McClellan instead ordered the retreat to continue, several of his senior subordinates became irate. Gen. Phil Kearny exploded, "Such an order can only be prompted by cowardice or treason!" As Union forces left the peninsula, northern journalists and political cartoonists had a field day with McClellan, portraying him as indecisive, cowardly, and childlike. For the most part, scholars have been equally unkind in their assessment of the Young Napoleon. Many military historians argue that McClellan should have launched his attack on Richmond sooner; that he should have ordered his own counterattack south of the Chickahominy or after Malvern Hill. Instead, Little Mac, in the words of his foremost biographer, Stephen Sears, "lost the courage to command" and continued the retreat a beaten man.[63]

That argument has much to recommend it. McClellan's failure to take Richmond owed a great deal to his excessive caution and near paranoia about Confederate troop strength. His decision to retreat from Lee instead of launching a counterattack also demonstrated a lack of flexibility or aggressiveness. Even as President Lincoln pondered relieving McClellan of his command and abandoning the peninsula, the general steadfastly refused to acknowledge any flaw in his original plan. Little Mac insisted that he could still succeed if provided with another 60,000 troops for a second assault on Richmond.[64]

McClellan's personality and questionable decisions make it difficult to evaluate the role of weather in the Union defeat. Comparisons using modern weather data suggest that overall the spring and summer of 1862 might have been only slightly wetter than usual. What seems more important is *how* the rains came to the peninsula that year, falling for several days or a week at a stretch in April and early May and again in late May and early June. It also rained frequently in late June and early July. The timing of the precipitation could hardly have been worse for McClellan. It delayed his advance at Yorktown and made the roads to Richmond nearly impassable; it flooded the Chickahominy and kept his army divided as it drew close to Richmond; and it turned the retreat to Harrison's Landing into a sickness-inducing slog.

The same rain fell on Confederate forces, too. However, Johnston and Lee had fewer men and animals to move on the muddy roads. The southerners also maintained essentially a defensive position for most of the campaign, occupying the higher ground around Richmond as their enemy slowly made its way through the swamps along the Chickahominy. Like their Union counterparts, Confederates suffered from dysentery and diarrhea, though the exact percentage of southern troops afflicted with the ailments is difficult to gauge. Having lived in the South, Rebel soldiers might also have had more acquired immunity to malaria, though not all the troops had grown up in regions where the disease was endemic. Had it not rained so much, would McClellan have led a relatively healthy army to capture Richmond and end the war before President Lincoln issued the Emancipation Proclamation? Such speculation is as dangerous as it is appealing. However, it remains impossible to divorce McClellan's actions from the natural environment in which they occurred.

As if by design, when McClellan's army reached Harrison's Landing, the rain that had plagued the campaign suddenly stopped. Temperatures soared on the peninsula as the southern summer set in. A week after Malvern Hill, McClellan wrote his wife, "The day is insufferably hot—intense—so much so that I have suspended all work on the part of the men." By August, even the usually loquacious general had run out of words to describe the sweltering conditions: "I can't convey the idea of the heat today . . . not a breath of air stirring."[65]

With a La Niña pattern in place in the Pacific and the unsettled weather of winter and spring a memory, the Civil War Drought tightened its grip on the West and South. In California, three years of barely measurable rainfall followed the devastating floods of the previous winter. By August, hot, dry conditions prevailed from the Gulf States north to the Ohio River and east to the Atlantic Seaboard. On July 8, a Massachusetts soldier occupying New Bern, North Carolina, near the southern tip of the Outer Banks, was astounded to find his "thermometer standing 108 degrees in the shade." A fellow soldier commented on the South's infernal summers, "I don't believe the devil would live here if he wasn't obliged to." Even the highest reaches of the southern Appalachians—regions that occasionally got eighty inches of precipitation a year—had become bone dry. Troops who depended on locating water as they marched across the South now found themselves woefully short of that precious resource, especially as the fighting on the war's western front spread into Tennessee and Kentucky.[66]

Rather than driving deeper into Mississippi after capturing Corinth at the end of May 1862, Union general Henry Halleck divided his forces and sent them after different objectives. Halleck ordered Gen. Don Carlos Buell to move the Army of the Ohio through northern Alabama to capture Chattanooga, Tennessee. The drought made Buell's excursion more difficult as the

Tennessee River, at flood stage during the winter and spring, fell so low by June that Union vessels could not travel upriver to supply Buell's army. Confederate cavalry raids also made the railroad lines unreliable as a source of supply. Union raiding forces that had preceded Buell's army that summer had already stripped most of the provisions from northern Alabama farms. As a result, Buell's parched troops had to live off half rations and whatever potable water they could find. Some veterans remembered the campaign as their hardest march of the war. The Army of the Ohio made it to Stevenson's Depot, a railroad stop thirty miles from Chattanooga, before they received orders to change direction and chase down a Confederate army moving into Kentucky.[67]

The Confederate force heading to the Bluegrass State was the Army of the Mississippi, led by Gen. Braxton Bragg. Bragg had replaced P. G. T. Beauregard as commander of the army after the Confederate evacuation of Corinth. After weighing his options to relieve pressure on Mississippi, Bragg decided to act on a plan suggested by Gen. Edmund Kirby Smith, who was operating out of east Tennessee. Smith recommended joining forces with the prospect of eventually invading Kentucky, where he was certain thousands of recruits would flock to the southern colors. In July 1862, Bragg moved his army eastward from Tupelo, Mississippi. He took a circuitous 776-mile railroad and ferry route through Mobile, Alabama, and arrived at Chattanooga, Tennessee, in early August. Bragg entered into a loosely coordinated arrangement with Smith to force the Union armies in Alabama and Mississippi to fall back to protect their supply lines in Tennessee and Kentucky. Deviating from the plan, Smith invaded Kentucky in mid-August, prompting Bragg to follow suit later in the month. After defeating a small Union force at Richmond, Kentucky, on August 30, Smith essentially played no further role in the campaign, resting his army near Lexington and not making serious efforts to unite with Bragg's army. However, the movement achieved one of its goals. The mere prospect of two southern armies in Kentucky, threatening Cincinnati and the major supply hub of Louisville, forced Buell to abandon his Chattanooga campaign and race north to confront the Confederate menace.[68]

Whatever the overall strategy, men in the ranks usually had one major objective: finding water. Hounded by Unionist guerrillas and baking in the extreme heat, Bragg's men grew faint as they marched through the Tennessee mountains in late August. Officers had to post guards around stagnant, putrid pools of water to keep away soldiers willing to risk sickness in order to slake their thirst. As they entered Kentucky, one Tennessee soldier recalled, "the only water accessible was pond water . . . so muddy that we could not wash our faces in it." Passing former Union camps, they discovered that many water sources were ruined as Federal troops had "butcher[ed] cattle and dumped offal" in the pools, "making the water unfit to drink." Bragg twice had to alter the path of his march into Kentucky because reports indicated no water or provisions along the planned route. As a result, he appeared to be driving directly for Louisville, heightening Union fears.[69]

When Bragg's army got within forty miles of Louisville, businesses closed, residents fled, and the commander of the city's garrison ordered all women and children to evacuate across the Ohio River to Indiana. Fortunately for the city's panicked residents, Buell's army got to Louisville first, at the end of September, worn thin from the arduous march across the parched landscape. They were, one observer noted, "the hungriest, raggedest, tiredest, dirtiest, lousiest and sleepiest set of men" in the United States. The men got only one week's rest before Buell, feeling pressure from the War Department to defeat the invaders, marched his troops southeast seeking Bragg's army.[70]

Bragg's men had set up a defensive position near Perryville, a small village located at a strategic crossroads near the center of the state. Perryville's citizens usually drew their water from the Chaplin River and its tributaries, but by autumn 1862, the streams had "shrunken into little, heated, tired-looking threads of water, brackish and disagreeable to taste and smell." Bragg's men had control of the few viable watering holes that remained. As the Union soldiers approached Perryville, one Indiana soldier recalled that the men were "almost frantic. . . . The topic of conversation was without exception of the one thing on all minds, water, water, water." A soldier from Illinois remembered finding only "nauseating" pools, "blue and thick from the slime of frogs and hogs." Cohesion began to break down in several Union units due to the desperate search for potable water. A Pennsylvania soldier reported, "More than once five dollars was offered and refused for a single drink of water." As the Union troops closed on Perryville, many soldiers slipped out of the ranks to hunt for water, only to be captured by the enemy. By the first week of October, they were more than willing to kill for a chance to drink from a full canteen.[71]

On October 7, Union troops tried to drive southern soldiers away from Doctor's Creek, a small stream now composed mostly of near-stagnant mud holes. The Yankees succeeded at first, but Rebel forces thought the water so valuable that they counterattacked on October 8 and the Battle of Perryville began in earnest. It was a confused fight, in which only a fraction of the respective armies participated. Even so, soldiers on both sides remembered Perryville as some of the fiercest fighting of the war. Though not defeated, Bragg—concerned that he was outnumbered and already dismayed by the paltry number of recruits and amount of supplies his army had obtained in Kentucky—decided to withdraw Confederate forces from the state. Union soldiers, however, likely cared less about the Rebel retreat than the turn in the weather. Two days after the battle, heavy rain saturated the region, replenishing—for the moment—the very resource that had triggered the confrontation.[72]

Accounting for the role of weather in war is not a novel concept. Nor was it entirely unknown in the mid-nineteenth century. Carl von Clausewitz, a Prussian military theorist, whose famous treatise, *On War*, was published posthumously in 1832, incorporated weather into his ideas about the ways "real war" differed "from war on paper." Under battlefield conditions, Clausewitz believed, victory or defeat might hinge on myriad unforeseen factors, including the deeds or decisions of a single person. Such influences, which Clausewitz termed "friction," could perhaps be explained, in retrospect, but they could never be anticipated. Clausewitz argued that weather played a similar role in war: "Rain can prevent a battalion from arriving, make another late by keeping it not three but eight hours on the march, ruin a cavalry charge by bogging down the horses in mud, etc."[73]

Clausewitz's concept of "nature as friction" has much in common with Edward Lorenz's ideas about the "butterfly effect," chaos theory, and nonlinear dynamics, concepts useful in explaining human interaction with the natural world. Simply put, people act and make decisions in "hybrid environments," created by "ever-changing fusions of human and non-human actors and activities." Commanders execute military strategies in real time on the physical battlefield where weather, microbes, soils, animals, and soldiers commingle to create a shifting environmental mosaic, one influenced as much by nature as by human nature. The true test of a general's merit, Clausewitz believed, lay in his ability to "know friction" and to adjust his tactics and expectations to accommodate and overcome the unexpected.[74]

Although Clausewitz's work did not appear in English until 1871 and was unavailable to American Civil War generals, it illustrates a fundamental truth about weather and war: acknowledging nature's agency in Civil War battles does not absolve military leaders of responsibility for their actions. Indeed, viewed in the context of a hybrid environment, command decisions become not only more nuanced and understandable but also more susceptible to criticism and interpretation, especially in highly variable conditions like those of 1862. Grant might be applauded for adapting to and taking advantage of flooding conditions along the Tennessee River, while McClellan's actions on the peninsula might well have cost the Union a victory that could have ended the war. Those respective adjustments—how each dealt with nature's friction—inevitably reflected the style, aggressiveness, and other personality traits of the generals, characteristics that would, in time, elevate Grant and get McClellan relieved of his command. On the southern side, one might point to Sibley's failure (despite his previous experience in New Mexico) to make adequate preparation for desert conditions, a source of friction that halted what had been a successful Confederate effort in the West. In contrast, Robert E. Lee used delays caused by wet weather to prepare for his offensives against McClellan, actions that helped drive Union forces away from Richmond and set the stage for Lee's rise to godlike status in the South. And at Perryville, it was water, or the lack thereof, that prompted Buell and Bragg to confront each other in one of the most intense battles of the war.

For most Americans who lived through it, the turbulent weather of 1862 had a far simpler explanation. They saw their personal fates and the outcome of battles as resting solely in the hands of God. In their eyes, God sent the rain that led to the fall of Forts Henry and Donelson. God sanctioned the storms and unusually wet weather that confounded McClellan, as well as the Union's forced withdrawal from Alabama. In a world devoid of chaos theory and even Clausewitz's ideas about friction, weather proved even more unpredictable and unknowable than it is today. But just like today, it continued to play a crucial role in everyday events, especially in an agricultural nation where the available food supply hinged on successful planting and harvest. As the conflict dragged on, the constant search for sustenance, brought on by capricious weather and the machinations of war, would influence dozens of military decisions and have profound implications for the civilian population. By the fall of 1862 many Americans on and off the battlefield already had more than a passing familiarity with food shortages. And things were about to get worse.[75]

Three

FOOD

FALL 1862–SUMMER 1863

Making a living from the soil has never been easy. By 1862, no one knew that better than the people of Culpeper County, Virginia. Situated in hill country where the Piedmont narrows to meet the Blue Ridge, the county had once produced tobacco in abundance. By the mid-nineteenth century, however, that demanding plant had so depleted the best lands that wheat and corn had emerged as the county's most viable crops, both for consumption and sale. Even those grains required the application of various fertilizers, crop rotation, and the systematic planting of legumes such as clover and peas to return nitrogen to worn out soil. Such practices required year-round work and in Culpeper County, as in much of the South, a lot of that labor came from enslaved African Americans. In 1860, well over half of the county's 12,000 residents were slaves.[1]

Economically, the system proved a huge success. When the war began, Culpeper was the thirteenth-most-prosperous county in Virginia (out of 148 counties). Ecologically speaking, however, it was a shaky and vulnerable system. It required enormous amounts of human energy and constant manipulation of nature to sustain it. An insect invasion, a sudden storm, or a host of other potentially devastating forces could jeopardize both subsistence and profit. Each day became an ecological tug-of-war, as county residents tried to take what they needed from the land without destroying it.[2]

For more than a year, President Lincoln had insisted that northern armies not add to the problems of farmers like those in Culpeper County. George B. McClellan and other northern generals had ordered their troops not to confiscate private property, including foodstuffs, as they moved through the state. Leaders hoped that this official policy of conciliation might convince southerners to look more favorably upon the Federal forces and perhaps rejoin the Union. But by the spring of 1862, with Confederate resolve stiffening instead of weakening, some northern commanders had already begun to ignore the president's orders. That April, near Luray, Virginia, an Illinois officer watched his men help themselves to "beef, mutton, ham, bacon, bread, honey, onions, chickens, ducks, geese, turkey, eggs, butter, bread [sic], milk, sausages," and other items in the lush valley. Evoking Old Testament imagery, he wrote that local residents would "always remember the advent of the 'Yankee soldiers' among them as the Egyptians do the flight of the Locusts."[3]

By July, Lincoln, too, began to rethink official policies on dealing with southern civilians. Conciliation had been predicated on the notion that Union troops would win a quick victory, a prospect that grew bleaker by the day during the summer of 1862. After Little Mac failed to take Richmond, Lincoln ordered him off the peninsula and instructed McClellan to join forces with Union general John C. Pope. Pope had won some minor victories along the Mississippi River that spring and now commanded the new 50,000-man Army of Virginia that protected Washington, D.C. As McClellan moved north, Robert E. Lee sent 25,000 Confederate troops to engage Pope's men. When Stonewall Jackson defeated an element of Pope's army at Cedar Run on August 9, Lee took his remaining 30,000 soldiers and set out to reunite his forces, now known as the Army of Northern Virginia.[4]

With Lee again on the move and precious few southerners returning to the Union fold, the Federal government decided to scrap the official "kid glove" policy toward civilians. Congress instituted new rules that allowed for confiscation and destruction of private property, including crops. Not all Union leaders agreed with the tactic, but General Pope seemed delighted with the new strategy. It was high time, he noted, that the secessionists, including the prosperous farmers of Culpeper County, felt "the hard hand of war."[5]

With an army four times larger than the number of civilians in the county, it did not take Pope and his men long to lay the landscape bare. Soldiers cleared out smokehouses, barns, and fields of grain. Troops took horses, wagons, and fence rails, most of which became fuel for Union campfires. One northern newspaper correspondent wrote, "Unoffending citizens have been impoverished in a day . . . their fencing destroyed, their sheep and hogs and cattle butchered, their grain entirely consumed." He declared, "Many a family has been left in a condition verging upon absolute want and starvation." With the army present, farmers were afraid to tend their crops; much of their remaining wheat and corn baked to death in the summer heat. After

Union soldiers taking farm animals from civilians in western Virginia, August 1861. Such foraging expeditions became more common beginning in the summer of 1862. From *Harper's Weekly*, August 17, 1861.

Pope's army left the county, one Confederate correspondent noted, "The fertile fields of Culpepper [*sic*] are now one desert waste." And more hungry soldiers were on the way.[6]

By the time Union forces began to strike directly at the southern food supply, Lee's soldiers had long been accustomed to living off the land and scavenging whatever victuals they could from the enemy. Stonewall Jackson's men had marched several hundred miles in the Shenandoah Valley campaign during the spring and fed frequently on Yankee supplies. Indeed, they had captured so much from Gen. Nathaniel Banks's army that they had dubbed that unwittingly generous general "Commissary" Banks. But when they moved to defend Richmond in late June, they fought hard on limited rations. Jackson did not perform well in the Seven Days—frequently arriving late and not demonstrating his usual aggressive tendencies. Historians have offered plenty of excuses for his poor performance, but some contemporaries suggested that Jackson had held his men back because they were

Food

exhausted and hungry. In the eight days leading up to and including the campaign, the commissary had issued Jackson's men only two days' rations of meat and six days' rations of hardtack. In August, outside Washington, D.C., they had better luck.[7]

In order to defeat Pope before McClellan joined him, Lee divided his army again, sending Jackson with 25,000 troops around Pope's northern flank and well to his rear. Jackson captured Pope's major supply depot at Manassas Junction on August 27, and his soldiers reveled in their good fortune. Boxcars and warehouses bursting with gastronomic delights tempted the hungry soldiers, who filled their haversacks with cakes, candy, fish, ham, lemons, oranges, wine, and canned goods of great variety. One soldier noted, "To see a starving man eating lobster salad and drinking Rhine wine, barefoot and in tatters, was curious. The whole thing was indescribable." The feast did not last long. The Confederates lacked enough wagons to transport the Union stores and — well aware that they would soon be famished again — the troops reluctantly put the torch to the supplies they could not carry. As Lee described it, "Quantities of stores had to be destroyed for want of transportation." Inadequate food distribution and the inability to take advantage of such windfalls from the enemy would plague southern forces throughout the war.[8]

Jackson's troops took cover in the woods west of Manassas. Pope found Jackson the next day and launched repeated attacks against him for two days near the old Bull Run battlefield. On August 30, while Pope committed his whole force to breaking Jackson's defensive line along an old railroad cut, Lee launched a devastating attack on the Union left flank with the recently arrived divisions under the command of Gen. James Longstreet. The defeated Union army retreated toward Washington the next day, having lost 16,000 men in the two-day fight — as many as the much larger Army of the Potomac had sustained in the weeklong battles near Richmond earlier that summer. Lee's army suffered 9,000 casualties, fewer than half the number he lost in the Seven Days.[9]

Recognizing the demoralization of the Union Army of Virginia, Lincoln relieved Pope of command and reluctantly placed McClellan in charge of the combined Union armies, acknowledging that, "if he can't fight himself, he excels in making others ready to fight." Meanwhile, Lee made what proved to be one of the most important strategic decisions of the war. Hoping to capitalize upon the recent Confederate successes, he elected to push his army — fatigued and famished but flush with victory — across the Potomac River into Maryland. He intended to drive deep into Pennsylvania, all the way to the state capital of Harrisburg if possible.[10]

Several factors influenced Lee's decision to move on Maryland and Pennsylvania. He wanted to press his advantage against the demoralized Union armies, encourage the growing peace sentiment in the North ahead of the midterm elections that November, and convince European nations (particularly Great Britain and France) to join the war on the South's behalf. However, before any of those objectives could be accomplished, Lee's men had a far more pressing need: food. The general knew that the inept southern commissary system could not provide adequate sustenance for troops on the move, and they had little chance of finding much in battle-scarred northern Virginia. Moreover, if the army remained, civilians in Culpeper and surrounding counties might well starve, too, as they competed with soldiers for what additional food that fall's meager harvest might bring. Those who saw Lee's army as it prepared to cross the Potomac that September struggled for words to describe the malnourished troops. "When I say they were hungry," noted one woman, "I convey no impression of the gaunt starvation that looked from their cavernous eyes." Gazing upon the men from afar, one observer wondered if scarecrows from all the fields of Maryland had somehow assembled in one place. "A most ragged, lean and hungry set of wolves," another called them.[11]

The invasion of the North was a huge gamble, but given the state of his troops, Lee felt he had little choice. "The army is not properly equipped for an invasion of an enemy's territory," he admitted to Davis on September 3. "It lacks much of the material of war, is feeble in transportation, the animals being much reduced, and the men are poorly provided with clothes, and in thousands of instances are destitute of shoes." Only the lush farms of western Maryland, still untouched by war, with barns, one soldier noted, "as large as Noah's Ark," offered the prospect of adequate nourishment. With such an apparent cornucopia on the other side of the Potomac, Lee decided to take the risk. "We cannot," he asserted, "afford to be idle."[12]

Maryland indeed had food, but to the Confederates' dismay much of it was not yet ready for harvest. Constantly on the move, Lee's soldiers grabbed anything at hand, most often green corn and green apples. For men on the verge of starvation and desperately in need of meat and bread, the steady intake of green fruit and vegetables did their digestive tracts more harm than good. Both foods, with their high fiber and sugar content, led to increased fluid in the large intestine and quick evacuation of the excess water through the bowels. After six days of such rations, Pvt. Alexander Hunter of Virginia explained, "There was not a man whose form had not caved in, and who had not a bad attack of diarrhea." Every soldier's "under-clothes were foul and hanging in strips." As the haggard army moved toward Frederick, Maryland, one resident marveled that he could smell the troops before he saw them. "I have never seen a mass of such filthy strong-smelling men," he told a reporter for the *Baltimore American*. "Three in a room would make it unbearable." Lee's men suffered from "galled crotches, bleeding and ulcerated rectums," as well as gullets sore from retching, symptoms that left the southern soldiers "weakened and irritable." The image of thousands of threadbare Confederate soldiers, reeking of excrement, bile, and body odor, squatting to defecate or bending over to vomit in the fields and woods lining the roads to Maryland lacks romance, but such were the hard realities of sustaining the southern army.[13]

At Frederick, Lee's men cleaned out the store shelves, but that briefly provided food for just a fraction of the troops. Many farmers had not yet gathered their wheat crop, and millers were reluctant to grind corn or wheat for the Confederates. Cattle were also not nearly as abundant as they had hoped and, as one Alabama soldier noted, hunger still "gnawed at our stomachs mightily." By mid-September, on the eve of one of the war's most important battles, many Confederates survived on a mere eight ears of corn a day, which provided fewer than 1,000 calories. Lee launched his invasion with 55,000 soldiers (including three new divisions that replaced the casualties of Second Bull Run), but within two weeks more than 10,000 of his soldiers had fallen out of the ranks. As Alexander Hunter noted, "The whole route was marked with a sick, lame, limping lot."[14]

Convinced that McClellan would move with his usual caution, Lee anticipated being in Pennsylvania before he had to deal with the Union army. Thus, he divided his soldiers into several groups, sending two-thirds of his brigades by different routes to capture Harpers Ferry, Virginia, at the confluence of the Potomac and Shenandoah Rivers, and moving the remainder to Hagerstown, Maryland, near the Pennsylvania border. Unbeknownst to Lee, however, an errant copy of his Special Orders No. 191, which detailed the marching routes for his divided army, accidentally fell into Union hands. Once aware of the Confederate strategy, McClellan put the Army of the Potomac in motion much faster than Lee expected. A small portion of Lee's army fought a delaying action at South Mountain on September 14, while Lee tried to unite his disparate forces. When he learned on September 15 that Jackson had captured Harpers Ferry, the Confederate general decided to concentrate his army at Sharpsburg, Maryland, only seventeen miles away. He loathed abandoning the invasion without a fight and remained confident he could defeat McClellan in a pitched battle.[15]

Situated in the Hagerstown Valley hard by the Potomac River, Sharpsburg was home to some 1,300 residents. Like farmers across the Blue Ridge in Culpeper County, most of those in and around Sharpsburg made their living cultivating grains and raising livestock. They, too, relied on various fertilizers and crop rotations to maintain their fields. Some residents with larger land holdings used slave labor, but the institution was nowhere near as pervasive as in northern Virginia—in 1862, 150 slaves lived in Sharpsburg, along with 203 free blacks. Like the once bountiful farms of Culpeper County, those around Sharpsburg never failed to impress visitors. One Union officer declared that the town and surrounding fields made for "as pleasing and prosperous a landscape as can easily be imagined." Still, the Marylanders knew the inherent risks in deriving sustenance from the soil. Before the arrival of the armies, the greatest hazard to the year's crop might have been the developing drought. The wet spring of 1862 had long since given way to a bone-dry summer. Clear days made it possible to work long hours in the fields, but the lack of rain threatened the maturing corn. However, for the rural people around Sharpsburg, as for those of Culpeper County, the biggest threat to their food supply came not from nature but people. In mid-September, 120,000 men descended on them to fight what became known as the Battle of Antietam, named for a creek that flowed nearby.[16]

On the morning of September 17, Confederate general John Bell Hood's men were in good spirits—not in anticipation of battle but because finally they had a chance to cook breakfast, the first bona fide meal they had eaten in three days. When the Battle of Antietam began at dawn, Hood ordered his men to the front before they could eat a bite. "Mad as hornets" because they had to abandon their breakfast, the soldiers made their way toward a lush thirty-acre cornfield owned by David R. Miller. There, in fighting so fierce that it turned a field of six-foot-high corn stalks into stubble, many of Hood's men died with their stomachs still empty. When the fighting at "The Cornfield" ended in a stalemate, the center of the action shifted to the "Sunken Road," and then to a bridge on Antietam Creek (now called "Burnside's Bridge" in honor of the Union general who fought his way across it that afternoon). When night fell on Sharpsburg, 4,500 Union and Confederate soldiers lay dead; another 18,000 had been wounded. Two days later Lee retreated to Virginia. Antietam proved a strategic Union victory, one that broke Lee's string of successes and helped turn the tide of war in favor of

Union troops charge through David Miller's cornfield on the north side of the Antietam battlefield. When the fighting ended, every corn stalk had been cut close to the ground as if by a scythe. From Johnson and Buel, *Battles and Leaders of the Civil War*, 2:630.

the North. Five days after the battle, Lincoln issued the preliminary Emancipation Proclamation, thereby changing the character of the war, making it about more than preserving the Union.[17]

Realizing that the lack of adequate food had contributed to the campaign's failure, the Confederates attempted some practical changes, instituting reforms within the commissary and supply bureaus to increase production and procurement of sustenance for the army. The Commissary Department reorganized and centralized its purchasing arrangements, while the government passed new legislation to provide more food for the armies. It took time for these reforms to take effect, and they never became very efficient. Lee's soldiers continued to go hungry for much of the war.[18]

For the people of western Maryland, the misery that began on September 17 lingered for months. The arrival of the two armies, along with other assorted camp followers and support personnel, instantly brought a hundredfold increase in human inhabitants. For two days that September, Sharpsburg's population exceeded that of Chicago, at the time the ninth-largest (and fastest-growing) city in the United States. After Lee's retreat, 50,000 Union troops remained in the area for a month and a half, along with thousands of wounded soldiers housed in farm dwellings and outbuildings that had been converted into hospitals. Even with most of the Confederates gone, more people lived within a five-mile radius of Sharpsburg than in Pittsburgh, Detroit, Milwaukee, Rochester, or Cleveland. Those cities, however, were established urban centers whose residents relied on food produced elsewhere—in the hinterlands of the American West and the Ohio Valley— to sustain them. The human horde that descended on western Maryland in 1862 created an instant metropolis that lacked the commercial ties and transportation networks to provide sustenance for its exponentially increased population.

Residents who returned after the battle quickly discovered that anything edible had vanished. Troops made off with jams, jellies, canned goods, flour, cornmeal, and still-green apples from the local orchards. One family came home to find that soldiers had killed dozens of chickens in their abandoned kitchen, leaving heads and entrails ankle-deep on the floor. Elsewhere around Sharpsburg farmers returned to find fences smashed and soldiers happily butchering hogs, cattle, and sheep. Thousands of stragglers from both armies who drifted into the region in the days after the battle only compounded the problem. By late September, a newspaper noted, Sharpsburg had "been eaten out of food of every description."[19]

Even with sustenance scarce, humans and animals in this new city produced tons of waste. Though the Sanitary Commission made it clear that Union soldiers "should never be allowed to void their excrement elsewhere than in the regularly established sinks [latrines]," enforcing the regulation in camp or on the battlefield proved impossible. Many men defecated and urinated where and when it suited them, with little regard for the health hazards their excrement created. On the Confederate side, the gastric distress among Lee's men made it even more difficult to attend to sanitation. One visitor who arrived at Antietam several days after the battle noted that he could still track the movements of Lee's men "by the thickly strewn belt of green corn husks and cobs" and the "ribbon of dysenteric stools just behind."[20]

Humans were not the only source of ordure. The daylong battle killed scores of animals, but as of October 1, Union forces still had 22,493 horses and 10,392 mules present. Modern estimates suggest that a healthy 1,000-pound horse or mule produces an average of 50 pounds of manure and six gallons of urine in a typical day. Those numbers can vary about 30 percent up or down, depending on an individual animal's food and water intake. Allowing for inadequate nourishment among war horses and taking the lowest possible estimate (35 pounds of manure and four gallons of urine per animal per day), Union horses and mules left at least 575 *tons* of solid waste and more than 130,000 *gallons* of liquid effluent on the landscape around Sharpsburg *every day*.[21]

Nineteenth-century Americans were no strangers to the hazards of such waste. Though they knew nothing of the bacteria that actually caused sickness, many physicians and public officials associated cholera and typhoid with the general filth of humans and animals living in crowded urban conditions. By the 1860s the sixteen largest American cities had moved toward "water carriage" systems of waste disposal. Those systems used water pumped from distant locales to carry waste out of the cities, usually into rivers and streams. These new sewer systems created their own set of environmental problems, but in the short run, city health improved.[22]

Sharpsburg had no comparable method of waste management. Most local folk probably relied on outhouses and privies and used manure from stables and streets to fertilize their fields. Moreover, dry summer conditions had left many streams running low, and without steady rain to flush them out, local water sources became especially susceptible to contamination. In the days immediately after the battle, residents worked day and night to bury corpses and amputated limbs in shallow graves until they could be properly interred in a cemetery. Soldiers covered dead horses with lamp oil and set fire to the carcasses. Even so, sickness plagued the region for months after the Union army left. Exact diseases and numbers of those afflicted are difficult to determine, but residents reported outbreaks of typhoid and chronic diarrhea; some (perhaps mistakenly) spoke of cholera. All those diseases resulted from crowded conditions and bacterial contamination of water and food supplies. The area around Sharpsburg retained a reputation for general sickness — most of which local people attributed to the aftermath of the battle — throughout the winter of 1862–63. Only after the spring rains, which helped flush out streams and replenish groundwater, did the most acute sickness abate. Some residents believed that the town did not recover until well after the war ended.[23]

With no meat in the smokehouse, no corn in the crib, no crop to sell, and no immediate government assistance to help them, Sharpsburg's civilians faced ruin. Farms went bankrupt and some families depended on charity to survive. Such was the aftermath of nearly every major battle of the war. Troops devoured every scrap of food when they arrived and left a near ecological wasteland when they departed. After Antietam, a resident noted, "You couldn't hear a dog bark nowhere. You couldn't hear a bird whistle or a crow caw.... We didn't even see a buzzard with all the stench.... When night come I was so lonesome that I see I didn't know what lonesome was before. It was a curious silent world."²⁴

The dry weather of late 1862 in Sharpsburg signaled the return of the decade-long drought. It was especially devastating because it followed an exceptionally wet spring. The same rain that stymied McClellan on the Virginia Peninsula created enormous difficulties for the planting season in the South. Wet weather swamped fields from Louisiana to Virginia. A Connecticut soldier ferrying on a steamboat between Baton Rouge and Vicksburg remarked on the inundated Mississippi River plain in late June 1862: "The general features are dreary in the Extreme as there have been terrible floods here which in most cases has Completely ruined the crops and drowned a great many Cattle." Indeed, the river levels of the Mississippi, Yazoo, and Ouachita rose higher than they had in nearly fifty years, breaching levees and ruining thousands of acres of productive lands. The same was true across the South, as the incessant rains throughout April and May left the land too wet to plow.²⁵

The heavy precipitation created ideal conditions for stem rust, a fungus that devastated southern wheat crops in 1862. In Virginia's Shenandoah Valley, wheat production fell 50 percent from the previous year. Overall estimates suggest that the South lost about one-sixth of its potential wheat harvest. Potatoes and oats also suffered from fungal infections due to wet weather. After noting that his wheat was "hardly worth cutting," Piedmont South Carolina planter David Golightly Harris echoed the frustration of many southern farmers. "Everything has got the rust," he complained. "I feel rusty myself."²⁶

Growing wheat in the marginal habitat of the southern uplands had always been a challenge, and most southerners did not rely on that grain to keep them fed. Planters knew that success in war, indeed their very survival, depended on one crop: corn. Corn and pork were dietary staples for southerners of every class. When Frederick Law Olmsted toured the South in the 1850s, he ate corn bread at every dining table from the coast to the mountains. Corn blades (the green husks around the cobs) provided the chief fodder for cattle. The South had nearly doubled its corn production from 30 million bushels in 1850 to 55 million in 1861. Well aware that war might bring food shortages, southern newspapers urged farmers to produce corn and other foodstuffs instead of traditional cash crops. As one Georgia editor put it, "Plant corn and be free, or plant cotton and be whipped."²⁷

However, as the wet spring of 1862 gave way to summer drought, southern farmers slowly realized that fickle weather could trump even the most patriotic intentions. After watching rust infect his wheat, David Golightly Harris had high hopes for his corn. But in early July, the rain stopped. "DRY, DRY, DRY," he wrote in September. "My corn crop is cut short by the unprecedented drought. I much fear I will not make enough to [feed] my family." In the southern Appalachians, early frosts compounded the problem. Mountain farmers in Mitchell County, North Carolina, worried that the "hard freezes that visited this country much earlier this season than heretofore" had left local people with less than "half enough grain to bread their own families" and none "to let their neighbors." By 1863, corn production in the Shenandoah Valley had fallen by 60 percent.[28]

Occasional bad harvests were nothing new. Unlike the soils of the Midwest and New England that had been churned up and deposited by glaciers during the last ice age, most southern soils had been exposed to the elements for millions of years. As a result, much of their original mineral content had been leached away by rain, even before humans began to farm the land. Lacking calcium and other buffering agents, the region's soils, especially along the Atlantic Seaboard, tended to be more acidic than in the North and Midwest. Demanding crops such as tobacco and cotton quickly used up the available nutrients, leaving behind even more acidic "sour" land that seemed, as many visitors noted, best suited to weeds and pine trees.[29]

As part of their ongoing negotiations with nature, many southerners practiced field rotation. They planted tobacco or cotton first, followed by less exhausting crops until yields declined. Farmers then cleared a new tract, allowing the old plot to lie fallow. In time, it might be returned to cultivation or left alone, eventually to regenerate forest. Economically, the system appeared to be a success. Cotton accounted for 50 percent of American exports after 1840, and, by 1860, 60 percent of the wealthiest Americans lived in the South. Ecologically, however, the system had two major flaws. First, it required much labor and manipulation of nature. Southerners of sufficient means got that labor from slaves; others relied on themselves and their kin. Second, the system demanded fresh lands for clearing and, ideally, periodic expansion onto new untilled acreage. When the federal government forcibly removed Indians from the Deep South in the 1830s, southerners streamed into Alabama and Mississippi, especially to the famed Black Belt region marked by soils much richer in minerals than the lands to the east. Some in the South favored other remedies. Virginia planter Edmund Ruffin won fame as an advocate of crop rotation and treatment of old fields with calcium-rich marl that "sweetened" the soil and retained its vitality. Several prominent politicians and even one U.S. president urged cotton growers to invest in guano (bird droppings harvested thousands of miles away on the South American coast) for use as fertilizer. Most southerners, however, still put their faith in labor and land. Not long after the war broke out, both began to disappear.[30]

About 75 percent of white southern men worked their own land with help from family and neighbors. So many of those farmers had enlisted during the martial excitement after secession that few remained to till the fields. Most of those who joined the southern army simply entrusted care of their farms to wives and children. While family members could handle day-to-day farm chores and business transactions, the heavy work of clearing new fields, spring planting, and fall harvest required male labor. As early as November 1861, some western North Carolina counties found that there were "not enough [men] left to plant corn."[31]

Things got worse on the home front in April 1862, when the Confederate Congress passed a conscription law, siphoning off even more white labor. Conscription officers rigorously enforced the new measure. In early 1863, eastern North Carolina families watched helplessly as Gen. D. H. Hill took men straight from the fields. As one of Hill's officers described it, "I took up 38 [men] and nearly all their families (generally large) are almost entirely destitute of food and the means to produce it. But military power is inexorable and [Hill] swayed it with a merciless hand." The conscription officers also took millers (who ground the corn into meal), tanners (who made the leather necessary to work the farm animals), and blacksmiths (who repaired farm implements and made shoes for draft animals). Such conscription often devastated local economies, making it difficult to process and sell any crops that survived the drought. Pleading with Governor Zebulon Vance to take notice of their plight, a group of North Carolina women explained, "Last Fall our Husbands brothers & sons were carried from their Farms into the Army, consequently there was very little small grain sowed." As a result, "famine is staring us in the face. There is nothing so heart rending to a Mother as to have her children crying around her for bread and she have none to give them." With men gone off to war, the lingering drought, once merely a seasonal anomaly, now brought the threat of starvation. Moreover, southern leaders knew that hungry people could quickly lose their appetite for fighting. As David Golightly Harris succinctly put it, crop failure "at this time would conquer us much sooner than the Yankey would do it."[32]

In peacetime, when supplies ran low, cattle and hogs had helped south-

The Confederate government passed the Conscription Act in April 1862, and army officials rigorously enforced it. The act added men to the armies but drained manpower from farms, cutting into the harvest of food crops.
Courtesy Library of Congress.

erners fend off famine. Slaughtered in winter, even a few head of poorly nourished half-feral livestock might see a family through a lean season. However, to preserve beef and pork, southerners needed salt, a staple now in critically short supply. By the summer of 1862, as the first crops began to fail from drought, encroaching Union troops choked off the usual salt supply from several sites in West Virginia. Union leaders there offered salt at a discounted price to encourage residents to defect from the Confederacy. For pragmatic farmers with "wives and large families of helpless children," Union offers of salt were "more powerful than armies."[33]

In time, the Federal blockade curtailed salt imports to many southern ports as well. Some southerners found creative ways to cope with the shortage. Those on the coast cooked rice, grits, and hominy in seawater. Others ripped floorboards out of salt houses and soaked them in water, before boiling it down to gather a few grains. They captured loose salt grains from cured meat and sifted the soil under smokehouses to find and reuse salt. Some threatened violence against Confederate authorities. In December 1862, a group of women in Greenville, Alabama, marched on a railroad station and confiscated a large supply of salt from a local agent while shouting "Salt or Blood." Farmers did have the option of keeping their animals alive through winter, slaughtering them one at a time, and selling any meat they could not immediately use. Without salt, however, livestock no longer provided as much insurance against hunger.[34]

Moreover, as 1863 wore on, the South's most valuable resource—land and the soil that covered it—increasingly fell into enemy hands. When the war began, southern farmers had nearly 57 million acres under cultivation. By January 1, 1863, Federal forces controlled all of Kentucky and two-thirds of Tennessee, where the Confederacy lost 4 million acres of viable farmland. Before the war, those states had also produced much of the corn and pork consumed in the South. Union troops captured another 4 million acres along the coasts of South Carolina, Georgia, and Louisiana—seriously curtailing the South's production of rice—as well as the coastal regions along North Carolina's Outer Banks and its large fishing industry. West Virginia, which had seceded from the government in Richmond in 1861, officially joined the United States in June 1863, taking away 2.3 million acres, while Union armies in northern Virginia rendered another million useless.[35]

Even as millions of acres once devoted to food crops disappeared, southerners continued to devote much of their cultivated land to cotton. Because the South supplied Great Britain with about 80 percent of its imported cotton, many southern leaders believed the British would eventually enter the war on the side of the Confederacy. In 1861, southerners instituted a cotton embargo in an attempt to compel them to do so. Confederate officials in several port cities formed committees of public safety to make sure that southern cotton stayed out of British hands. The tactic backfired spectacularly. A worldwide cotton surplus kept British textile mills running for a year, and by mid-1862, Great Britain had found enough alternate sources of cotton in India and Egypt to weather the storm. Meanwhile, southern fields that might have been devoted to corn, field peas, and other subsistence crops had been sacrificed to King Cotton.[36]

With food disappearing and no sign that Britain might enter the war, southern civilians had little choice but to cut back consumption. Some found new and creative uses for old staples. Coffee, an imported item, disappeared soon after the war began. Kate Stone, living in Louisiana across the river from Vicksburg, remarked in March 1863 that "after experimenting with parched potatoes, parched pindars, burned meal, [and] roasted acorns, all our coffee drinkers decided on okra seed as the best substitute." Others simply had to do without. A western North Carolina historian noted that during the war provisions were so scarce for his mother and siblings that the family only ate two meals a day instead of the usual three. They became so conditioned to the wartime regime that "after the war was over and provisions had become plentiful they never really enjoyed the third meal again."[37]

In the spring of 1863, with children hungry and no relief forthcoming, some southern women resorted to violence. In Atlanta, on March 18, 1863, a woman entered a store and brandished a pistol at the owner while her accomplices helped themselves to bacon and other goods. They told onlookers that "their suffering condition" required such desperate measures. Atlanta was only one of a dozen Georgia cities that experienced food riots that spring. In Randolph County, Alabama, farmers wrote to Jefferson Davis of a series of riots in which women seized government-owned wheat and corn "to prevent Starvation of themselves & Families." Over three weeks in March and April, similar riots over food, primarily bread, occurred in Mobile, Alabama; Salisbury, North Carolina; and Petersburg and Richmond, Virginia.[38]

The most famous of these food riots was the one in the Confederate capital. On April 2, a crowd of nearly 1,000 women, primarily wives of Tredegar Iron Works employees as well as working women from a clothing factory, marched through Richmond's streets toward the governor's mansion. When War Department clerk John B. Jones asked what was going on, "a young woman, seemingly emaciated, but yet with a smile, answered that they were going to find something to eat." Armed with various weapons, they ransacked shops in the nearby commercial district. Confederate soldiers stopped them at the city market, where President Davis climbed on a wagon and ordered the women to disperse or be fired upon. Within days, the Richmond City Council began a program to provide more food for the poor. Across the South, threats of similar action forced local governments to institute similar relief programs. As a result of the food riots, historian Stephanie McCurry explains, "The states of the C.S.A. would build a welfare system on a scale that was unprecedented in the history of the South or the North in terms of budget and administrative commitment."[39]

Even such government largesse could not alleviate the triple environmental threat of bad weather, lost labor, and diminishing land. As the Confederacy's armies continued to go hungry, its government passed legislation that unwittingly made the situation even worse for civilians. On March 26, 1863, the Confederate Congress enacted an impressment law, allowing the armies to seize food (which they had, in fact, been doing since 1861) and pay a price set by the government for the seized goods. Designed to help alleviate the shortages within the army and urban populations, the measure actually had the opposite effect. The law deterred farmers from growing food crops because they knew their harvests would be impressed and the government would likely pay only half as much as the produce was worth. Several state governors and various newspaper editors argued that the policy would starve both soldiers and civilians. Commissary Gen. Lucius Northrop thought otherwise. Commenting on the dwindling grain supply, he offered a simple and chilling assessment: "The army must have what there is, and the people must go without."[40]

A month later, the Confederate Congress followed up with the tax-in-kind law. Based upon the church tithe system, the new measure allowed farmers to reserve 50 bushels of sweet potatoes, 100 bushels of corn or 50 bushels of wheat, and 20 bushels of peas or beans for their own use, but it demanded 10 percent of all remaining produce above those minimums. The tax engendered some protests, but nothing on the scale of those that followed impressment. The new policies, however, often fell victim to the crumbling Confederate logistical system. Tons of food supplies collected by impressment agents in the southern hinterland rotted for lack of adequate transportation or shelter. As a result, agents focused on citizens living closest to the armies, picking nearby lands clean while farmers far removed from the fighting gave up less. Agents also resorted to stopping loaded trains and wagons bound for urban areas and redirecting them to the armies, creating near-famine conditions in cities like Richmond.[41]

In the Confederate capital (as in other southern cities), prices for scarce provisions soared. When War Department clerk John B. Jones noticed the price of basic food items rising dramatically in October 1862 (with flour at $16 a barrel and bacon at 75 cents a pound), he wondered, "How shall we subsist this winter?" By March 1863, flour sold for $40 a barrel and even rancid bacon brought $1.50 a pound. In July, Jones noted, "We are in a half starving condition. I have lost twenty pounds, and my wife and children are emaciated to some extent." On January 1, 1864, flour prices had more than tripled to $150 a barrel. By early June, a single barrel cost $500. Prices dropped to $200 per barrel in August but quickly rose again. By the beginning of 1865, merchants could get $700 a barrel, and on March 20, 1865, just three weeks before Lee's army surrendered, wealthy citizens desperate for bread paid up to $1,500 a barrel! Jones and his son had a combined income of $8,000 a year — a very good living by nineteenth-century standards but not nearly enough to buy bread at those prices. Most of the city's residents were far less fortunate than the Jones family.[42]

All the while, soldiers from both armies continued to ravage the southern landscape in search of food. In the spring of 1863, Union cavalry under Gen. Grenville Dodge attacked supply depots in northern Alabama, destroying 1.5 million bushels of corn and 1 million pounds of bacon, enough to keep nearly 45,000 soldiers on full rations for a month. A Massachusetts soldier near Union-occupied New Bern noted that war had left the countryside so bare that it was now "a desert as hopeless as [the] Sahara itself." Outside Richmond, an area subject to nearly constant scavenging by southern troops protecting the capital, one woman had few kind words for the men fighting for her freedom. They "broke down my hen house and took every single chicken and all my ducks," she explained. "And they burnt up the fence and pulled down the garden and stole the Irish potatoes and my peas and my raspberries and my cherries and they trod down the cabbage plants. I don't know what we are to do nor what will become of us." Like Virginians before him, North Carolina governor Zebulon Vance turned to the Old Testament to explain the starvation and destruction he witnessed. "If God Almighty had yet in store another plague worse than all others which he intended to have let loose on the Egyptians in case Pharaoh still hardened his heart," Vance intoned, "I am sure it must have been a regiment or so of half-armed, half-disciplined Confederate cavalry."[43]

Even those who lived far from major battle sites sometimes had to deal with foraging soldiers. In remote regions, partisan guerrilla fighters often waged irregular war against Union or Confederate forces. From the North Carolina coast to the Appalachian Mountains to the hinterlands of Arkansas and Missouri, renegade soldiers took food from local residents regardless of their loyalties. Southern citizens also suffered from retaliation when guerrillas attacked Union forces. In December 1862, after a guerrilla unit attacked a Federal naval convoy on the Mississippi River, near Helena, Arkansas, Union forces burned nearly every farm for fifty miles along the river, destroying the residents' winter food supplies. One southerner bemoaned that in the spring of 1863, "those guerrilla bands of ours that swarm around Memphis I think caused those people who lived around there an immense deal of trouble." They stole food from locals, which, he argued, did "just about as much harm as the enemy could possibly do." A woman in western Missouri reported that "our farms are all burned up, fences gone, crops destroyed, and no one escapes the ravages of one party or the other."[44]

While southern farmers struggled against labor shortages, land lost to the enemy, and hungry soldiers who swarmed like locusts, production on northern farms exploded. On the sometimes thin but generally fertile soils left by retreating glaciers, New England farmers had turned to a system based on grass and large animals. The cool, wet climate made it possible to give over much land to permanent meadows. Cattle and sheep then turned grass into manure that could be collected in barns and spread on fields to restore and maintain fertility. Hogs foraged in woodlands that also provided timber for construction and firewood. Wheat, rye, oats, and barley (grasses that had, over centuries, been adapted to human use) grew in fertilized fields with fences to keep out wandering livestock.[45]

Farmers in Pennsylvania and the Mid-Atlantic colonies quickly adapted their system to those regions, where rich soil allowed for better grain harvests. By the 1860s, beef, dairy products, and grain not only kept northerners fed but also found new markets in the Northeast's emerging cities. Even so, mobility sometimes proved more viable than management. As population pressure increased, northern farmers and newly arriving immigrants increasingly sought subsistence and profit in the fertile, untilled soil available in the Midwest.[46]

With a population of 22 million compared to the 9 million people living in the seceded states, the North could better withstand the labor lost when some of its men went to war. In addition, new technology helped compensate for lost manpower. John Deere's 1837 patent on a self-polishing steel plow made breaking new ground much easier. Cyrus McCormick's mechanical reaper, conceived in 1831 in the Shenandoah Valley before McCormick moved his manufacturing center to Chicago, and Obed Hussey's invention of a horse-drawn reaper, made harvesting wheat and other grains more efficient. At nearly the same time, John and Hiram Pitts of Maine invented a mechanized threshing and fanning machine that allowed for the quick and efficient processing of the harvested grain. As a result, agriculture spread through the prairies of the Upper Midwest, the new technology helping farmers make the most of heretofore untilled soils. By the time the war began, an average family farm in the Midwest yielded enough surplus crops to feed 8.8 adults or about two additional families. The long-settled farms of New England did not produce as much—slightly less than enough to feed one additional person—but by 1860, most northern farms "were not only able to meet their own consumption needs, but also produced a surplus that could be marketed."[47]

Buoyed by the burgeoning midwestern market economy, northern agriculture continued to boom throughout the war. Despite the South's celebrated reputation as an agricultural society, the North had three times as many farms, with twice the number of cultivated acres, worth more than two and a half times the total cash value, than all of the seceded states combined. Growth continued even during the war years. Between 1861 and 1865, some 430,000 new farms sprang up in the region, bringing an additional 2.7 mil-

Cartoon showing a starving Confederate soldier and the abundant food available on the other side of the Ohio River. Shortages of food prompted Confederate incursions into the North. From *Harper's Weekly*, September 20, 1862.

lion acres of land under cultivation. In the midst of war, the Union doubled the amount of wheat, corn, beef, and pork it shipped overseas. In 1862, even as southern armies drove McClellan off the peninsula and moved on Maryland, Great Britain imported 45 percent of its wheat from the United States.[48]

Advances in food preservation allowed the Union army to take advantage of the agricultural bonanza. The canning process had been developed before the war, but preserved food proved expensive, and many consumers feared that it might not be safe. Isaac Winslow, a canner in Maine, pioneered the use of metal cans to preserve sweet corn, a development that helped reduce the risk of contamination. Beginning in 1860, Isaac Solomon of Baltimore perfected methods for speeding up sterilization and introduced use of the pressure cooker, a device that increased the quality of canned goods and lowered their cost. Condensed milk, a product invented by Gail Borden, had been available before the war but not widely distributed. That changed in the fall of 1861 when a Union commissary agent began purchasing large quantities of it for soldiers.[49]

To help move the food, the North employed an extensive distribution system consisting of thousands of miles of macadamized, all-weather roads, and nearly 4,000 miles of canals. In 1860, the United States also had 31,000 miles of railroads—nearly as much as the rest of the world combined. Over 22,000 miles of those roads were in the northern states, but in the Union army's advance into Mississippi, they utilized significant portions of the southern railways in Tennessee and Mississippi. In Louisville, Kentucky, a supply depot for the North's western armies produced 3,000 rations a day in 1861. By 1864, production of rations at the depot had increased more than a hundredfold, and Louisville was only one of many such depots that soon made the Union army the best-fed fighting force the world had ever seen. Confederate soldiers, accustomed to eating whatever they could find, frequently marveled at both the quantity and quality of northern rations. Upon entering an abandoned Union campsite in Virginia, one southern boy seemed almost awestruck: "The whole country around here is bright with tin cans used by the Yanks for vegetables, condensed milk, lobster, oysters, fruit & everything else."[50]

As the North began to build up an enormous surplus of food, southern armies continued to starve. By the winter of 1863, severe food shortages had begun to restrict tactical options for Confederate military leaders. In January, after not receiving adequate supplies from the Commissary Department in Richmond, Lee ordered his army's quartermaster to take fifty wagons and collect all the wheat possible between Richmond and Fredericksburg because "our necessities make it imperative that every exertion be made to supply the army with bread." His letters for the next three months contained dire warnings about the weakness of the soldiers and horses in his command due to insufficient food and forage. The soldiers agreed, as one Georgia sergeant mordantly quipped, "The Yankees say that we have a new gen'l in command of our army & say his name is *General Starvation* & I think for once they are about right." Lacking fruit and vegetables, Lee's men began to suffer from scurvy. In desperation, the general ordered each regiment to detail men to search daily for "sassafras buds, wild onions, lamb's quarter, & poke sprouts." Though they did contain vitamin C, the effect of this diet was limited because "for so large an army the supply obtained is very small."[51]

When Lee learned that one Union corps had moved by sea to Norfolk, Virginia, in February, he dispatched Gen. James Longstreet with two divisions of infantry and two battalions of artillery—20,000 men, or roughly one-quarter of his army—to southeastern Virginia not only to counter the threat but also to gather much-needed food for the rest of the army from a part of the state not already stripped bare. As a foraging expedition, Longstreet's venture was a rousing success. His men commandeered every wagon and ox cart they could find, collected tens of thousands of bushels of corn, and impressed nearly 1 million pounds of bacon from the coastal regions on either side of the North Carolina–Virginia border. Claiming that "it will require a month to haul out the supplies," Longstreet was delayed in rejoining Lee's army late that spring, but he had managed to gather enough provisions to feed Lee's army for nearly two months.[52]

The Army of Northern Virginia desperately needed sustenance. On March 27, Lee noted that his troops had survived for weeks on only eighteen ounces of flour, four ounces of "indifferent quality" bacon (one-third of the normal meat ration), and less than an ounce of rice a day. A South Carolina private wrote, "Sometimes we don't have more than three little slices of bacon for six men." Often soldiers ate only one meal a day or devoured their three days' cooked rations in two. Lee rightly worried that his soldiers "will be unable to endure the hardships of the approaching campaign." In early May, soon after the Union army's spring offensive began, several southern soldiers immediately surrendered themselves to advancing Federal troops at Fredericksburg, who explained, "They came in to get something to eat; they were tired of fighting on an empty stomach."[53]

In their weakened state, Lee's army faced a replenished, reinforced, and refreshed Army of the Potomac under a new commander. President Lincoln relieved General Burnside of command after the ill-fated Mud March in January 1863 and named Gen. Joseph Hooker as the army's new chief. An aggressive leader, Hooker quickly developed a brilliant plan to envelop Lee's army on both flanks. His offensive was so swift and surprising that Lee did not have time to recall Longstreet's troops from southeastern Virginia. Leaving 40,000 troops at Fredericksburg to occupy Confederate troops, Hooker prepared 70,000 troops to cross the Rapidan and Rappahannock Rivers to the west and attack Lee from the rear. On May 1, Lee learned of the movement and divided his 60,000-man army—leaving 15,000 at Fredericksburg and taking 45,000 to meet Hooker's flanking threat. The two forces met at a tiny crossroads village known as Chancellorsville.[54]

Though outnumbering Lee, Hooker inexplicably took a defensive position, awaiting Lee's next move. After a cavalry scout reported that Hooker's right flank was unprotected, Lee divided his small force again and sent Stonewall Jackson with nearly 30,000 troops on a circuitous flanking march to attack the exposed flank. Using the thick woods as a blind, Jackson's men burst out of the forest upon the unsuspecting Union soldiers late on the afternoon of May 2 and rolled up Hooker's right flank for nearly two miles. As darkness fell, Jackson fell victim to friendly fire and suffered a grievous wound; he died eight days later. After continued heavy fighting at Chancellorsville and Fredericksburg on May 3, Hooker ordered the Union army to retreat across the rivers, abandoning the campaign. The Union army had suffered 17,000 casualties, while Lee had lost not only Jackson but 13,000 men in his victory.[55]

Despite the astounding accomplishment given the odds he faced, Lee felt the victory would be meaningless unless he could follow it up with an even bolder stroke. While President Jefferson Davis favored sending part of Lee's army to Mississippi to help drive the Union forces out of that state, Lee suggested a far more daring plan—another attempted invasion of Pennsylvania. A victory there might compel the Union army to abandon northern Virginia, embolden Peace Democrats, and weaken Republicans, while reopening the possibility of foreign intervention on behalf of the Confederacy. But Lee also had a more immediate objective. The rich farms of Pennsylvania would feed his starving army and replenish depleted commissary stores. Buoyed by Lee's recent success, Davis approved the plan.[56]

As the reinforced Army of Northern Virginia crossed into Pennsylvania in late June, southern soldiers filled their letters with commentaries on the agricultural abundance that was such a change from the scarcity they had seen in Virginia. One Virginia cavalryman called Pennsylvania "the finest wheat country I ever saw before." "Big barns, & fat cattle, & fruits & vegetables were every where," recalled an admiring E. Porter Alexander. Though Lee issued orders against pillaging property, South Carolina soldier Tally Simpson declared that "the soldiers paid no more attention to them than they would to the cries of a screech owl." He noted how they "steal and pilfer in the most sinful manner" taking "any and every thing they can lay their hands upon."[57]

Some officers completely ignored Lee's directive. Gen. John Bell Hood announced to his troops, "Boys, you are now on the enemy's soil, stack your arms and do pretty much as you please." Even Lee seemed to turn a blind eye to the pillaging. Simpson wrote of how one day the army commander rode by while three dozen soldiers grabbed all the fowl on a farm as the lady of the house denounced them in strong language. When she saw General Lee riding by, she angrily called to him, "I wish to speak with you, sir." According to Simpson, "The Genl, without turning the direction of his head, politely raised his hand to his hat and said, 'Good morning, madam,' and then went his way." Soldiers helped themselves to all the grains, tubers, legumes, chickens, honey, apple butter, eggs, milk, and meat that the farms had to offer. It was a high time for the famished southern soldiers throughout the last week of June—but the revelry would end abruptly on July 1 at Gettysburg.[58]

While Lee's army feasted in Pennsylvania, Union troops 1,000 miles away at Vicksburg, Mississippi, turned food—or more specifically, the lack of it—into a weapon. After capturing New Orleans in April 1862, Union gunboats moved methodically up the Mississippi River, capturing Baton Rouge, Louisiana, and Natchez, Mississippi, without firing a shot. Northern strategists also intended to take Vicksburg, a crucial transfer point for food and supplies flowing into the Confederacy from the west. Jefferson Davis proclaimed, "Vicksburg is the nail head that holds the South's two halves together." Lincoln agreed. "Vicksburg is the key," he wrote. "The war can never be brought to a close until that key is in our pocket."[59]

Founded in 1819 by a cotton planter, Vicksburg was a flourishing commercial river port of 8,000 residents. Rows of docks and warehouses lined the waterfront, while many shops, three hotels, several churches, a synagogue, a convent, and dozens of magnificent homes dominated the hills above the river. The city's thriving economy depended on the cotton trade. Vicksburg served as the county seat of Warren County, home to nearly 200 cotton plantations, including one belonging to the Confederate president. Over the previous decade, county planters had doubled their cotton production to 14 million pounds in 1860, thanks to the labor of 13,763 slaves, out of a total population of 20,696 residents.[60]

Moreover, local geography rendered the city nearly impregnable. Perched on a bluff 300 feet above a hairpin bend in the Mississippi River and bristling with cannon, Vicksburg's garrison could fire with impunity on any Federal boats, which had difficulty navigating the river's bend and could not elevate their own guns high enough to do much damage to the city. In the summer of 1862, the city took on enormous symbolic importance as the last major Confederate bastion on the Mississippi. Union general William Sherman wrote in March 1863, "No place on earth is favored by nature with natural defenses as Vicksburg." For all its economic, geographic, and military advantages, however, Vicksburg had one glaring environmental weakness. As cotton production soared, the county's production of nearly every food crop had declined to the point that several planters purchased meat and corn to feed their slaves. Though farmers finally began planting more corn when

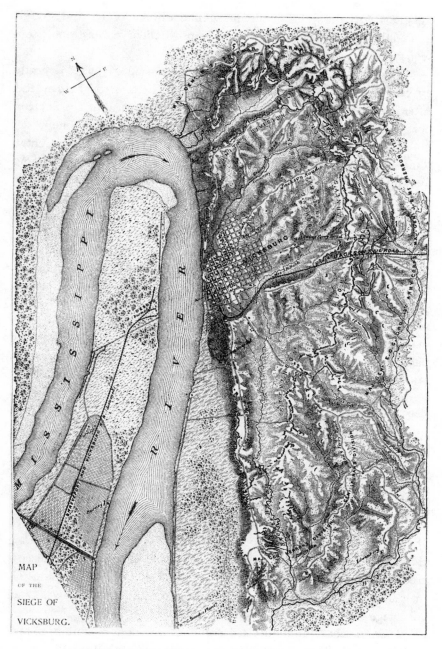

Map of Vicksburg and surrounding region. Gen. Ulysses S. Grant surrounded the city and starved the defenders into submission in July 1863. From Johnson and Buel, *Battles and Leaders of the Civil War*, 3:494.

the war started, the spring floods and summer drought of 1862 had severely limited their yields. The dearth of food crops would have catastrophic consequences in the summer of 1863.[61]

When the Union fleet retreated after a failed assault in June 1862, northern generals developed a plan to capture the Confederate stronghold by land. After turning back Rebel attacks in northern Mississippi at Iuka on September 19 and Corinth on October 3–4, Gen. Ulysses S. Grant—the commander of the Department of Tennessee—decided to follow the Mississippi Central Railroad south to Jackson, the state capital. From there he planned to move forty miles due west to Vicksburg. A substantial force under Sherman would move by boat from Memphis to the Yazoo River north of Vicksburg, securing Grant's right flank and tying down Confederate forces. Grant began his campaign at Grand Junction, Tennessee, on November 2, marching down the railroad line and quickly capturing Holly Springs, Mississippi, where he established a major supply depot. As the army advanced, Mississippi residents soon felt the harsher Federal confiscation policy. Union soldiers appropriated food, firewood, and other items from local farms. One Indiana soldier noted, "We lived high . . . taking from the inhabitants every thing in the eating line we wanted." Though Grant knew that foraging could partially sustain his men, horses, and draft animals, he planned to employ wagon trains to carry the majority of the army's food.[62]

As the Army of the Tennessee advanced deeper into Mississippi, Confederate forces under the command of Gen. John C. Pemberton seemed only too willing to retreat without a fight. In mid-December, however, Gen. Earl Van Dorn led a cavalry attack against Holly Springs and destroyed a mountain of food and ammunition that Grant had amassed there. Unprepared to live solely off the land, Grant withdrew to Tennessee. During the retreat, the Union general sent all his spare wagons to collect food and forage within a fifteen-mile radius of the road to Grand Junction. He "was amazed at the quantity of supplies the county afforded." The bounty collected during the retreat changed Grant's mind about the viability of foraging as he discovered that the Mississippi countryside could indeed keep his men and animals alive. As he later explained, "This taught me a lesson which was taken advantage of later in the campaign."[63]

As winter set in, Grant devised a series of ambitious operations to take Vicksburg, including rerouting the "Father of Waters" away from Vicksburg, thus negating the city's strategic importance. But each attempt to alter nature's preferred path failed. In April 1863, Grant decided to cross the river south of the city and approach it from that direction. It was a bold move, because Grant would have no dependable supply line for much of the campaign; his 44,000-man army would have to feed itself from the countryside. William Sherman wrote to his wife, "I look upon the Whole thing as one of the most hazardous & desperate moves of this or any war," and declared, "I have no faith in the whole plan." If Grant encountered unexpected Confederate resistance, unusually severe weather, or found insufficient food on southern farms, his army might face disaster. Despite Sherman's misgivings, Grant pushed ahead. Union vessels under Adm. David Porter steamed through the river obstructions and artillery gauntlet at Vicksburg on April 16, while the army marched south on the west side of the river. Two weeks later, Grant ferried his army across the river.[64]

Union forces quickly established a large supply depot at Bruinsburg, Mississippi, but officers warned soldiers to consider those rations for emergency purposes only. For daily subsistence they could rely on the lush farms nearby. The troops found more than they needed. As Sherman explained in early May, "In this starving country we find an abundance of corn, hogs, cattle sheep and Poultry." He quipped, "Men who came in advance have drawn but 2 days rations" in the previous 10 days "and are fat." Over the next three weeks, Grant's men marched 180 miles, fought five battles, and drew only five days' rations from the commissary. Mississippi farms supplied the rest.[65]

Other Union forces also lived off the land that spring. In order to befuddle Pemberton, Grant ordered Col. Benjamin Grierson to lead a large cavalry raid through central Mississippi. Grierson's primary objective was to destroy the Southern Mississippi Railroad east of Jackson, which would cut off supplies to Vicksburg from that direction, but the cavalry was also to destroy as much infrastructure and as many supplies as possible. Grierson and his 1,700 men set off from Lagrange, Tennessee, at dawn on April 17. Frequently dividing his fast-moving force into multiple detachments, Grierson created maximum confusion among Mississippi's military leaders and diverted as many as 38,000 Confederate troops, including most of Pemberton's cavalry, to chase the phantom forces.[66]

Grierson's soldiers ate well during those maneuvers. At New Albany, Mississippi, one woman bitterly complained that "my house, garden & orchard, were thronged with [Yankees] all the time, toating off corn & fodder, chickens, vegetables, [and] cooking utensils. They took every thing they could find that we had to eat." After tearing up the railroad and torching thirty-nine rail cars, Grierson's force rode west, linking up with Grant south of Vicksburg on May 2. Grierson had marched 600 miles in sixteen days, destroyed sixty miles of railroad track, tons of military equipment, and immense quantities of food while suffering only twenty-four casualties. Sherman called it "the most brilliant expedition of the war."[67]

Once his army crossed the river, Grant moved against Jackson, Mississippi, and drove away Gen. Joseph E. Johnston's small army of 6,000 troops massing there. After burning much of the capital on May 14, Grant turned west toward Vicksburg. Pemberton's troops fought two delaying actions before finally moving into the defensive fortifications that ringed Vicksburg's land side on the evening of May 17. As Grant's army approached, Pemberton ordered the city's noncombatants to leave, but approximately 4,500 chose to remain. When Mary Loughborough begged admittance into the city, saying, "We would meet any evil cheerfully in Vicksburg where our friends were," officials relented, but they ominously warned her, "If trouble comes, you must meet it with your eyes open." The arrival of Pemberton's 30,000 soldiers (and thousands of army animals) transformed Vicksburg into yet another instant metropolis. With a population of nearly 35,000, Vicksburg suddenly had as many people as Richmond, all of them squeezed into the town and garrison above the river. Sensing an impending crisis, several Warren County planters proposed that Pemberton take meat and corn from their larders and move them to the city. For reasons still unclear, Pemberton refused to heed their advice, a decision that sealed the fate of Vicksburg and its citizens.[68]

After two unsuccessful assaults against the city on May 19 and 22, Grant shifted his strategy. He decided simply to starve Vicksburg into submission. Union artillery pounded the city for the next six weeks, destroying dwellings, killing and maiming civilians, soldiers, and animals. In the first few days of the siege, artillery fire killed nearly 1,800 head of cattle within Confederate lines, immediately accelerating the garrison's food crisis as most of the meat rotted before it could be salvaged. In an effort to escape the bombardment, many residents took shelter in caves they dug into the clay bluffs facing the river. Pemberton hoped that Johnston would cooperate with him to help defeat Grant's army, or at least allow the garrison to escape, but in their correspondence over the next few weeks, it became clear that Johnston was not going to rescue Vicksburg.[69]

Those who had witnessed the devastation at Sharpsburg a year earlier could have predicted what Vicksburg faced. Prices for provisions soared as everyone—rich and poor, black and white—struggled against mounting hunger. By June 15, flour sold for an astounding $5 a pound (or $1,000 per barrel), cornmeal was $140 per bushel, and butchers scavenged the carcasses of animals killed by enemy shells to sell beef at $2.50 a pound. A fire burned through much of the city's business district on June 2, destroying nearly all the food remaining in stores. Any variety in diet quickly ended. Loughborough wrote that her family lived on corn bread and bacon for all three meals each day, "the only luxury of the meal consisting in its warmth." Another woman confessed, "I am so tired of corn bread, which I never liked, that I eat it with tears in my eyes." Before long, that bland staple began to vanish, along with flour, forcing Confederate commissary officers to search for substitutes from which to make bread. One such concoction, made from ground cowpeas, proved a miserable failure. No matter how long it baked, the meal in the center remained soft and raw, while the outer edges became so teeth-breaking hard, a soldier noted, "that one might have knocked down a full-grown steer with a chunk of it."[70]

After the first five days of the siege, the military garrison had to make do on one-quarter rations. Four ounces of flour and three ounces of bacon, with occasional tablespoons of peas or rice, provided the soldiers with no more than 900 calories a day. A diet with fewer than 1,200 daily calories compels the human body to begin devouring itself to provide the necessary energy to regulate normal bodily functions. Once the body has burned through all its fat reserves, it begins consuming muscles, which creates the look of emaciation. On June 4, a Louisiana soldier wrote, "the gaunt skeleton of starvation commenced to appear among the ranks." Eleven days later, a chaplain chronicled the slow, steady effects of starvation rations: "The cheeks became thin, the eyes hollow, and the flesh began to disappear from the body and limbs," making soldiers appear "haggard and careworn." Malnourished soldiers spent day after day in the trenches, where "the hot sun burned and blistered them," while diseases like malaria, "low fevers and dysenteric complaints" further weakened their bodies and their resolve.[71]

As at Sharpsburg, potable water disappeared. Vicksburg's wells and cisterns, which now had to supply five times as much drinking water as before, went dry. The weather did not help. No rain fell for a month after the siege began, leaving both residents and the landscape parched. Soldiers dug multiple wells, while civilians had to "buy water by the bucketful and serve it out in rations." Confederate troops also carried water from the Mississippi River to the defensive lines to relieve those who guarded the city, but the water "being conveyed so great a distance, became extremely warm and disagreeable." Additionally, drinking from the warm river, which carried human waste and dead animals, invited a host of diseases. By June 18, Union artillerists had zeroed in on places where soldiers fetched water, interdicting much of the water hauling.[72]

A fetid stench slowly settled on the city. Human waste overwhelmed privies and latrines. No one bathed for weeks. Dead mules, killed by Yankee shells or starvation, rotted in the sun even as some desperate soldiers tried to slice off and cook whatever putrid flesh remained on the carcasses. By the end of the siege, army officers issued orders to slaughter the "finest and fattest mules within the lines," in order to feed the troops. Hungry civilians and soldiers turned to any meat source they could obtain, including rats, which vendors dressed and sold in the city market for one dollar apiece. According to one observer, Vicksburg's residents consumed rats "in such numbers that ere the termination of the siege, they actually became a scarcity." "The thought of such food may be actually nauseating to many," admitted one Louisiana soldier, "yet, let starvation with its skeleton form visit them, and all qualms would speedily vanish." By the end of June, the city's inhabitants had eaten nearly every morsel of food available, including "fowls, cattle, horses, dogs, cats, frogs, or any living thing in reach except the gaunt human creatures who stared at one another with blood shot eyes and parched lips."[73]

By June 28, soldiers received only "one small biscuit and one or two mouthfuls of bacon per day." It was "not enough," they declared, "to keep soul and body together, much less to stand the hardships we are called upon to stand." The army had reached the breaking point, and a group of soldiers took the unusual step of writing directly to Pemberton: "If you can't feed us, you had better surrender us, horrible as the idea is." They warned, "This army is now ripe for mutiny, unless it can be fed." The malnourishment had taken such a toll that 6,000 soldiers now lay in the hospital, and many of the remaining troops could not muster the strength or the will to perform even the most mundane duties. One southern general asserted, "I do not believe that one-half of the men in the trenches could have stood up and fought had they been attacked by the enemy."[74]

With his men too weak to break out and no external relief forthcoming, Pemberton finally surrendered the city and its garrison on July 4. It was such a dark day in Vicksburg memory that the city did not celebrate Independence Day again until 1945. When the army capitulated, the commissary depot had no cornmeal or flour remaining, and only 38,241 pounds of bacon—barely enough to give each of the 30,000 men in the army quarter rations of bacon for six more days. Stunned by what they saw, Union soldiers took pity on their adversaries, offering food and assistance. "They were a starved, dirty, squalid set of rascals," wrote one Illinois soldier. Federal forces gave the inhabitants five days of rations each to alleviate their hunger. The fall of Vicksburg also had enormous consequences for food supply beyond the local region, as it cut the states of Texas, Arkansas, and Louisiana—as well as their 7.3 million acres of cultivated land and immense numbers of livestock—off from the rest of the Confederacy. By the end of that hot summer of 1863, the Confederate nation had lost fully one-third of its improved land (nearly 19 million acres out of 57 million), creating significant shortfalls in food production.[75]

When placed in historical perspective, though it mounted a valiant defense, Vicksburg did not hold out long. With a population of 35,000, it held out only forty-seven days, certainly nothing equivalent to some of the great sieges of military history—Jerusalem held out against the Romans for six months before succumbing in 70 A.D.; Antwerp held out for over a year against the Spanish in the late 1500s before capitulating; and Leningrad outlasted the Nazis by hanging on for 900 days in World War II. Strategic errors, however, undermined the Vicksburg defenders. Military officials had failed to stockpile food in preparation for a siege they must have known was likely. Moreover, food was in short supply in the cotton-centered country nearest the city. Ultimately, the South's agricultural focus on cash crops instead of food weakened its ability to prolong the fight and the war, and never was that more apparent than at Vicksburg in the summer of 1863.[76]

By then, people were not the only organisms in need of sustenance. Millions of animals played vital roles in the war effort on both sides. In the nineteenth century, farmers did not plant and armies did not move without horses and mules. Civilians and soldiers did not survive without protein provided by cattle and hogs. In much the same way as it affected human sustenance, the war put tremendous strain on animals and the forage that kept them moving. Managing animals became a problem for both sides and, as with the human food supply, the North's ability to keep animals fed and in the field proved crucial to the eventual outcome of the war.

Four
ANIMALS
SUMMER 1863–SPRING 1864

Life had not been especially kind to Lydia Leister. Widowed in 1859, she and her six children scratched out a living on a nine-acre farm just off the Taneytown Road, near Cemetery Ridge, about half a mile from Gettysburg, Pennsylvania. Planted in wheat, oats, and hay with small stands of apple and peach trees, the Leister property included a garden and a two-room house where the family had lived since 1861. Two years later, with the Army of Northern Virginia advancing on the town, Leister abandoned the farm for the comparative safety of a house along the Baltimore Pike. When she returned shortly after the Federal victory at the Battle of Gettysburg, Leister found that, in her absence, Union general George Gordon Meade had made her tiny dwelling his headquarters.[1]

The fighting around the farm had flattened Leister's wheat field. Soldiers took all the available hay from her farm, tore up her garden and orchards, and burned her fences for firewood. But those problems — by now expected after large engagements — paled in comparison to her biggest worry: what to do about the seventeen horses lying dead in her yard. The animals had been gathered near the house as Confederate artillery pounded Cemetery Ridge in preparation for the ill-fated southern assault on July 3 known as Pickett's Charge. When Rebel gunners overshot their targets, the shells fell on the Leister place. Meade and his officers escaped harm. Their animals did not.[2]

The seventeen carcasses on Leister's farm were some of the estimated 5,000 horses and mules that perished at Gettysburg. But those were not the only nonhuman casualties. At Cashtown, a small community where Lee's troops bivouacked before the battle, dead horses lay among "broken fences, slaughtered cattle, and other discarded martial debris." Touring the battlefield shortly after the smoke cleared, one observer saw as many as fifty dead oxen, noting that the animals "lie thick wherever the fighting was hot." Others reported all manner of dead cattle, chickens, sheep, and farm stock that had once belonged to civilians. Many of those animals had been butchered and half-eaten by soldiers.[3]

With so much carrion—not to mention more than 7,000 human corpses—local people expected the battlefield to be overrun with vultures, crows, and feral dogs. From all accounts, however, many of nature's scavengers, especially birds, had been driven from Gettysburg by the smoke and noise of battle. In the absence of those animals, scores of hogs from nearby farms rushed to fill the vacant ecological niche. Freed from pens torn apart by the battle and long accustomed to devouring anything humans left behind, the swine were soon "reveling in the [animal and human] remains in a manner horrible to contemplate." Occasionally the animals went after the living as well. At least one Union soldier watched as a scavenging hog tried to make a meal of his mangled leg. Other wounded men left on the field overnight had to use swords to drive away pigs foraging among the dead and dying.[4]

Many of those who visited the battlefield that July recorded similarly macabre scenes that shock modern sensibilities. Such matter-of-fact appraisals bear witness to the earthy quality of nineteenth-century life and the essential role of animals in nearly every human endeavor, including war. Like Lydia Leister, local people fretted about how to remove tons of decaying animal flesh, preferably without destroying their orchards and few remaining crops. Worries about foraging hogs focused not on the animals' scavenging habits but on the need for proper burials and treatment of the wounded, before the swine did exactly what rural folk expected.

Horses, mules, and oxen moved every supply wagon, artillery piece, and ambulance employed at Gettysburg. Troops on horseback were indispensable to both sides. Cattle, hogs, and even chickens moved alongside soldiers and munitions, providing a mobile meat supply for hungry men. When they had time, soldiers hunted deer, rabbits, and other wild game. That animals figured so prominently in the ecological aftermath of Gettysburg should come as no surprise. It was impossible to wage war without them.

The Gettysburg campaign began with a clash among horsemen. On June 9, just as Lee was preparing to move his army north toward Pennsylvania, the largest cavalry battle in North American history took place at Brandy Station, south of the Rappahannock River, in northern Virginia. Gen. J. E. B.

Dead horses from the Ninth Massachusetts Battery at Gettysburg. Lydia Leister faced much the same scene when she returned home to find seventeen dead horses in her yard. Courtesy Library of Congress.

Stuart's 9,500 cavalry troops camped near the station, screening Lee's army from prying Union eyes and masking his offensive intentions. At dawn, 11,000 Union cavalry troops under Gen. Alfred Pleasonton splashed across the river and surprised Stuart's force with a two-pronged attack. Over the course of fourteen hours, southern troopers finally forced the Yankees to withdraw. The southern cavalry lost more than 500 men, while the North suffered 900 casualties. Hundreds of horses on both sides fell in the battle. Stuart held the ground, but the Union cavalry gained an enormous morale boost as a result of its performance. Having suffered many defeats at the hands of Confederate horsemen before, the northern troopers felt they had finally matched their enemy on the field. Stuart's aide, Henry McClellan (a nephew of the Union's George McClellan), acknowledged, "One result of incalculable importance certainly did follow this battle—it *made* the Federal Cavalry."[5]

For Lee's army, however, the battle did not bode so well. From the early days of the war, Confederate troops had made do with far fewer horses than their Union counterparts. For two years, southerners had been able to offset that disadvantage with their superior horsemanship and better understanding of cavalry tactics. After Brandy Station, as the northern cavalry came into its own, Lee realized that the South would have to find more equine resources soon or his army and his country's bid for independence had little chance of success. The pressing need for horses as well as provisions to feed them played an important role in Lee's decision to move into Pennsylvania again.

Next to soldiers, horses and mules were the most important sources of energy on any Civil War battlefield. When the war began, neither side could have imagined the extraordinary number of horses that would be needed. The prewar U.S. Army only fielded about 15,000 horses, the vast majority serving in the nation's five cavalry regiments spread across the western frontier. Most military thinkers could not envision either a long war or large armies. Yet when George McClellan marched up the Virginia peninsula in the spring of 1862 with more than 110,000 men, he had 46,000 horses and mules with him—more draft animals than any previous army on North American soil had ever had soldiers. The numbers would only grow larger in the following years.[6]

Like their human counterparts, the animals first had to be pulled from civilian life and made ready for battle. In 1861, the United States was home to 1.1 million mules and 6 million horses—with 3.4 million in the North, 1.7 million in the South, and nearly 1 million in the border states. Private individuals owned nearly all those animals. The country had no state or national breeding farms that might guarantee a steady supply of animals, nor could such farms be established once the war began. Horses require an eleven-month gestation period and usually an additional four years to mature into an animal strong enough for military service; indeed, both armies preferred to use horses between four and nine years old. Even if a breeding farm had been started the day that Fort Sumter surrendered, the first horses would not have been ready for military service until February 1866, ten months after the conflict ended.[7]

Horses acquired for the army did not fare well early in the war. Lazy or unskilled soldiers ruined untold thousands of their mounts. While European armies dedicated three years to training cavalry soldiers, American soldiers were lucky to receive a few weeks of training. A former Dutch cavalry officer pointedly asked, "What, for instance, can be expected from a stocking manufacturer, or a linen weaver who considers the horse a wild beast?" He complained, "They never learn to ride, never can preserve their balance, but hang on the horse like a senseless lump, which in order to preserve its equilibrium, unnecessarily wastes a large portion of its strength, and on this account is soon exhausted." French observers estimated that Federal cavalry troopers were using up between three and six horses per man each year.[8]

Because of their superior animals and horsemanship, Confederate troopers had an early advantage. One northern official asserted, "Almost every southerner learns to ride while a mere child, and is perfectly at home in the saddle long before he arrives at manhood." Additionally, the official argued, they had been raised in a culture of thoroughbred racing, so the horses were better: "Every planter of means plumed himself upon his stables, and no southern town of any magnitude was without its race-course." While these northern perceptions did not necessarily match reality — not every southern horse was a thoroughbred, nor every southern boy a legend in the saddle — they point to important qualitative differences in troopers and horses.[9]

Union army officials also quickly discovered that their horse supply was grossly inadequate for an active cavalry. When it became clear the war would not end quickly, one observer noted, "It was then that the loyal men of the north were deeply mortified at the discovery that they possessed neither horses nor riders worthy of sustaining the glory of a legitimate cavalry service." In a wartime report to the commissioner of agriculture, Francis Morris stressed that it was vital that the country set up a program to raise a "good and true stock of horses." Col. Samuel Ringwalt agreed, declaring that "the common breeds of northern horses were sadly deficient" for use as cavalry. He argued, "We had uniformly associated the idea of speed with trotting, and neglected the easy gaits which alone can give the cavalryman the steadiness and command over his weapons which are essential in war."[10]

Because both sides had to rely on whatever animals they could muster from the home front, the South, with qualitatively better but quantitatively many fewer horses and mules from which to draw, faced an immediate problem. When the war began, southern cavalrymen had to provide their own horses. As a result, cavalry troops were on average wealthier than their infantry or artillery counterparts, because they had to have sufficient means to provide their own mounts. Self-interested soldiers frequently abused this system. An inspector of field transport for the Army of Tennessee grumbled,

"When a soldier is dismounted he is entitled to a furlough of thirty days to go home and remount himself. This makes [troopers] mere horse traders, selling their animals whenever they desire to go home." Increasing numbers of those men who left on horse furloughs never returned to the army. In the early going, southern forces sometimes captured horses and mules from the enemy, either taking the animals in battle or when Federal forces abandoned their exhausted mounts. Beginning in the spring of 1862, however, Union leaders ordered field commanders to kill any horses and mules their men left behind so the animals would not fall into Confederate hands. Applied sporadically at first, by 1863, the policy had begun to have the desired effect.[11]

With not nearly enough horses available for purchase on the open market, the Confederate government began to take animals from civilians. Though agents ranged to the far corners of the South in search of suitable mounts, southerners residing near army encampments suffered most from the policy. As a result, Virginia and North Carolina provided the greatest number of horses for Lee's army. Government impressment of horses had serious implications for civilian farmers. The lack of draft animals made it increasingly difficult to produce crops, making the food shortage worse for both soldiers and civilians. In time, desperate Confederate officials investigated acquiring mules from Mexico, swimming horses from Texas across the Mississippi River, and even shipping cotton and tobacco through enemy lines in exchange for the required animals. None of the measures succeeded. Looking back on the Confederate loss, at least one southern War Department official believed the lack of equines had been a major reason for the defeat, the "country [having been] stripped by impressments of horses."[12]

The Union system for procuring horses proved far more efficient, thanks to the managerial skills of Union quartermaster general Montgomery C. Meigs. Soon after the war began, the Federal government contracted with various sellers who guaranteed a specific number of serviceable horses at a set price, usually between $120 and $145 a head; by the end of the war, some cavalry and artillery horses were selling for $185 a head. Early on, those animals, and other horses purchased on open markets in larger cities, went to one of three major "procurement depots," located in Philadelphia, New York, and Cincinnati. By late 1863, large depots in St. Louis and at Giesboro Point in Washington, D.C., also helped to house horses and mules. Located on the eastern bank of the Potomac just below its confluence with the Anacostia River (the current site of Bolling Air Force Base), the sprawling 625-acre Giesboro Depot could accommodate up to 30,000 animals at a time. Over the course of Giesboro's existence, 170,654 cavalry horses matriculated

Giesboro Depot outside of Washington, D.C. Over 170,000 horses passed through the 625-acre facility during the war. Courtesy Library of Congress.

through the facility. Other smaller depots at Greenville, Louisiana; Harrisburg, Pennsylvania; Nashville, Tennessee; and Wilmington, Delaware, aided in getting horses and mules from suppliers to the battlefields. These distribution centers also allowed for rest and rehabilitation of animals exhausted in war. The recovery system became so effective that near the end of the war, Meigs estimated that "50 per cent of the horses which reach the depots disabled and broken down are returned ultimately to military service."[13]

Supplying horses to the Union army could be a lucrative business. One Cincinnati contractor claimed to have sold the government 50,000 animals worth $4.5 million by 1863. Reliance on such large suppliers was efficient, but it raised the ire of small farmers and country merchants who hoped to sell a few horses and mules to the military. In response to criticism that the process favored the wealthy, Union quartermasters turned more to animals bought on the open market in 1864 and 1865. For the most part, however, the energy that moved the northern army came from those with the financial wherewithal and business connections to deliver horses and mules in large numbers. With such a huge purchasing operation, charges of fraud in the contract process surfaced throughout the war. Corrupt inspectors passed off many horses younger than three or older than twenty years of age. Less than 50 percent of one shipment of New York horses was healthy enough for service. Gen. William Sooy Smith, a Union cavalry chief in the western theater, charged that "not one fifth" of the cavalry horses corralled at Nashville were serviceable. "Many of them are from fifteen to twenty years old," he complained, "some are blind and some are badly spavined [broken down]."[14]

Despite problems with the undemocratic system, the demand for horses remained insatiable. Between April and October 1863, Meigs purchased roughly 35,000 animals. Throughout the spring and summer of 1864, he was able to provide two remounts for every cavalryman in the Army of the Potomac, and, by the end of the war, the quartermaster general supplied 500 replacement horses a day to the army. Over the course of the war, the government procured approximately 1 million animals for transport and cavalry, some 650,000 horses and half as many mules. The total number of Confederate animals is more difficult to estimate, but by all accounts, southern forces made do with far fewer horses and mules.[15]

Stories abound of officers' close relationships with their horses. McClellan often spoke highly of his favorite mount, Daniel Webster, who survived the war and lived out his days in luxurious retirement on the family farm. Perhaps the most celebrated Civil War horse, Traveller, is buried near the general he carried into battle, Robert E. Lee. Though such emotional bonds were common among officers and cavalrymen, quartermasters and others charged with procuring animals for combat probably thought of horses less like companions and more like living machines, a view shared by most nineteenth-century Americans. The term "horsepower," apparently coined by James Watt when he invented the steam engine in 1775, derived from the work done by dray horses, a breed established to move heavy loads in English factories. In antebellum America, horses not only supplied power for mills but also were the engines for machines such as water pumps, sawmills, and construction equipment. In larger cities, horses pulled streetcars, canal boats, omnibuses, and fire engines.[16]

Like all machines, horses and mules required fuel, usually in the form of hay and grain. To remain healthy, an average 1,000-pound horse needed fourteen pounds of hay and twelve pounds of grain per day, while mules required three fewer pounds of grain. Hay supplied nutrition and bulk that supported the equine digestive system. Grain provided the necessary energy and nutrients for muscles. Horses also needed salt, preferably at least three times a week. The demand for such fuel could be staggering. An average cavalry force of 1,000 animals needed seven tons of hay and six tons of grain *every day*. Large herds could quickly and easily devour all the available grass around an army encampment. To keep the animals fit for battle, their rations had to be transported along with other supplies and delivered to them at the front.[17]

As in most matters of supply, Union methods of providing animal feed proved superior to those of the Confederacy. Using hay "presses" to compact palatable grasses into 400-pound bales, Union suppliers quickly became adept at cheap and efficient bulk hay shipments. Most forage for northern animals moved by rail, but Union quartermasters also shipped such provisions by water to the various supply depots. From there, either railroads or wagons moved the forage to the front. It was a huge undertaking. Northern horse rations took up over half the railroad cars and wagon trains supplying the Union armies. Occasionally temporary shortages occurred when key transport hubs fell into Confederate hands. For example, Gen. John Pope's cavalry horses suffered from a lack of forage when Stonewall Jackson's force destroyed the Union depot at Manassas in August 1862. Such setbacks notwithstanding, Union horses generally had adequate food throughout much of the war.[18]

Lacking comparable organization and infrastructure, Confederate armies often fed horses the same way they fed men, by allowing them to forage while on the move. During early 1863, Lee repeatedly complained about the chronic shortage of provisions for man and beast. "I am willing to starve myself," he wrote, "but cannot bear my men or horses to be pinched. I fear many of the latter will die." His quartermasters sent wagons as far as seventy miles away from camp to acquire forage, but the teams were frequently too weak to haul the loads back over the muddy winter roads. When Lee detached James Longstreet with two divisions to southeastern Virginia in February 1863, it was as much to provide forage for the army's starving animals as food for the soldiers. Lee dispersed thousands of cavalry and artillery horses to remote areas in Virginia and North Carolina in search of viable grazing land. As late as April 16, 1863, Lee described his army as still "very much scattered," as various units tried to keep themselves and their animals fed.[19]

Relocating half-starved horses to areas removed from the front created as many problems as it solved. When the animals could not find grass, cavalry officers began taking civilian grain for feed. Confederate agents even went after corn, a crop often in short supply and one much better suited to sustaining humans than horses. Worried about such activities in northwestern North Carolina in the spring of 1863, Governor Zebulon Vance asked that the horses be removed. As he noted, "the country was almost ruined by drouth [sic] last season [and] there will be the greatest difficulty in feeding the wives and children of absent Soldiers." When Secretary of War James Seddon balked at Vance's request, the angry governor did not mince words: "When the question is narrowed down to women & children on the one hand and some worthless cavalry horses on the other, I can have no difficulty in making a choice."[20]

As a result of critics like Vance, the Confederate military moved to create its own horse depots after the Battle of Brandy Station. Records of the number of horses kept at those locales are sparse, but the southern distribution system never rivaled that of the Union in efficiency. In the fall of 1863, the War Department created four inspection districts throughout the Confederacy to allow for the recovery of weak horses. The largest Rebel depot at Lynchburg, Virginia, housed just over 6,800 horses and mules during one fifteen-month period. Only about 1,000 of those animals ever returned to the front. As the war progressed, the lack of forage prevented the Confederate army from contesting Federal cavalry raids toward Richmond.[21]

In addition to fuel, equine machines required maintenance and spare parts, including saddles, blankets, bridles, bits, harnesses, and other tack. When it came to keeping the animals in action, however, nothing proved more important than horseshoes. Horses' hooves regenerate naturally, but the growth rate varies in each animal. Continuous work causes hooves to wear down faster than they can be replaced. Without horseshoes for protection, horses easily come up lame. During the first half of the nineteenth century, blacksmiths made most horseshoes by hand. In 1857, however, Henry Burden, president of New York's Troy Iron and Nail Works, developed a machine that allowed for mass production of a standard horseshoe. Burden's invention meant that horses could be shod almost anywhere with only a few simple portable tools. The U.S. War Department eventually spent over $3.3 million on shoes for Civil War horses, much of which went to purchase Burden's products.[22]

Meanwhile, Confederate quartermasters struggled to find shoes for their animals. With imports of suitable metals curtailed by the Federal blockade, the South had to develop internal sources. Lead mines at Wytheville, Virginia, and a smaller operation near Davidson, North Carolina, provided just over forty tons a month. The Confederacy also began mining iron ore in Alabama, laying the foundation for the postwar iron industry in Birmingham. As the shortage of metal became more acute, government agents scoured abandoned battlefields and appropriated piping, window weights, cistern linings, utensils, and other superfluous materials from civilians. Even if adequate materials had been available, the South lacked facilities for industrial production, as few smelting plants or ironworks existed in the region. Over the course of the war, individual blacksmiths still made more than half of all Confederate horseshoes.[23]

Fuel and maintenance notwithstanding, horses and mules differed from conventional machinery in important ways. As living beings, equines faced constant threats from other organisms, including those that cause disease. Gathering animals from distant locales, like the mustering of men, created nearly ideal conditions for the propagation of potentially dangerous microbes. Crowded together at supply depots, corrals, hayricks, and battlefield encampments, animals that had once lived in relative isolation from various contagions became hosts for bacteria and viruses brought to the front by other equines. Like soldiers, animals also suffered from unsanitary conditions and poor nutrition, making them even more prone to infection. Shortly after the Battle of Antietam, an outbreak of an especially problematic ailment known as "greased heel" infected more than 4,000 cavalry and artillery horses in the Army of the Potomac.[24]

Greased or "greasy" heel is a skin inflammation that usually shows up on a horse's lower legs. It can be caused by a variety of microorganisms and parasites, including several fungi, bacteria, and even mites or other insects. Without treatment, animals develop serious lesions that exude pus or a clear liquid, hence the "grease" in the ailment's name. Swelling of the limbs follows, and eventually horses go lame. It most often occurs either when horses work for long periods in wet weather or remain in close quarters where urine and excrement create muddy conditions. By early November 1862, Union artillery officer Charles Wainwright noted that the advancing disease had caused the hooves of the horses in his battery to crack, "in some cases so badly that you can put your finger between it and the crown." In the worst instances, Wainwright observed, "the hoof is almost off." Occasionally the disease "attacked in more than one foot" and rendered horses unable to stand. The ailment also put entire cavalry regiments out of action and might help explain McClellan's much criticized failure to move his army quickly after Antietam. When McClellan wrote to the president that he could not put his army in motion in October 1862 because his cavalry horses were tired, Lincoln facetiously responded, "Will you pardon me for asking what the horses of your army have done since the battle of Antietam to fatigue anything?" Perhaps greased heel was a contributing factor to the general's caution. However, the disease played no favorites. Lee's cavalry also suffered from greased heel during and after his retreat from western Maryland.[25]

Though debilitating, greased heel could be cured with rest and better attention to sanitation. That was not the case for the war's two most feared equine diseases: an incurable respiratory infection commonly called glanders and a disorder of the lymphatic glands known as farcy. Both resulted from infection by bacteria called *Pseudomonas mallei*, an organism easily passed between animals via shared forage grounds, watering troughs, and even grooming brushes. As glanders, the disease attacked a horse's respiratory system, creating ulcers in the nasal cavities and a thick discharge that, at its worst, could suffocate the animal. As farcy, the malady often affected glands on a horse's hind legs, causing them to swell and emit yellowish liquid. Both forms of the infection could kill and neither could be cured, but glanders nearly always proved more serious. An animal diagnosed with glanders usually had to be shot to prevent further infection within the herd.[26]

Glanders apparently made its first appearance early in 1862, when Confederate general Joseph E. Johnston retreated from Manassas just before McClellan embarked on the Peninsula Campaign. Though the source of initial contagion remains unclear, each side blamed the other for starting it. Union colonel Samuel Ringwalt went so far as to accuse the Confederates of biological warfare, noting that the Rebels had intentionally left behind sick animals to infect Yankee herds. The disease, however, probably originated the previous winter as both sides brought in new horses in preparation for the upcoming campaigns. Usually the armies had neither time nor money to construct separate stalls for their animals. In camp, horses remained staked out, tethered, or on pickets so they could be easily fed and quickly outfitted for battle or supply trains.[27]

Under such conditions, the glanders outbreak quickly became a wide-ranging and deadly epizootic. The contagion killed some 248,000 Union horses between 1861 and 1863. The disease might have been even more devastating among Confederate animals. Though records of horse mortality are neither as plentiful nor as detailed as those from the North, an estimated 75 percent of the South's military horses died every year. An outbreak of glanders at the Lynchburg Depot may well explain why so few animals initially housed there ever made it to the front. Neither side had an advantage when it came to dealing with glanders. No treatments—other than quarantine, less crowded conditions, and killing affected animals—could stem the contagion once it got started.[28]

Adequate veterinary care was practically nonexistent. With not a single veterinary school in the United States at the time, officers in northern and southern armies continually decried the "deficiency of veterinary talent in the country." During four years of war, the Federal government spent roughly $100 million to acquire horses and mules for the armies, but less than one-tenth of 1 percent of that total ($93,834.97) on medical personnel to treat them. Perhaps the lack of money allocated to caring for the animals reflected the government's view of horses as replaceable machines. One Confederate quartermaster estimated that the average artillery horse lived only 7.5 months, and cavalry horses likely fared worse. Union cavalry officer Charles Francis Adams asserted, "I do my best for my horses and am sorry for them; but all war is cruel and it is my business to bring every man I can into the presence of the enemy, and so make war short. So I have but one rule, a horse must go until he can't be spurred any further, and then the rider must get another horse as soon as he can seize one." Considering that only 10 percent of all their deaths resulted from battle, horses were far more likely to fall to bacteria, microbes, or neglect by their masters than to enemy fire. By the time the armies met in July 1863, the North's access to more horses, better forage, and mass-produced horseshoes had begun to affect the course of the war.[29]

After the Battle of Gettysburg, Lee marshaled his remaining horses in order to extricate his army from Pennsylvania. He put his supply wagons and ambulances in motion the night of July 3 as the army began a twelve-day trek back to Virginia. Confederate cavalry and artillery soldiers took every available horse from the Pennsylvania countryside as the massive caravan moved south. Southern artillery officer Robert Stiles admitted that "quartermasters, especially of artillery battalions, were, confessedly and of malice aforethought, horse thieves." Though local farmers tried to conceal their animals, Stiles noted that the soldiers "became veritable sleuth-hounds in running down a horse, and were up to all the tricks and dodges devised to throw us off the track." In addition, the Confederate quartermaster called for 8,000 to 10,000 horses to replace those killed or incapacitated in the campaign, but other than those scavenged during the retreat few were forthcoming.[30]

The Union army, too, took civilian stock in the aftermath of Gettysburg. Their quartermaster immediately called for 2,000 cavalry mounts and 1,500 artillery horses to get the northern force back up to standard. In response, the secretary of war ordered impressment of all serviceable horses in the region. That emergency policy, however, was short-lived because Union quartermaster Meigs had 7,000 replacement horses from various supply depots on the way to the Army of the Potomac by July 7.[31]

As the heat and drought of the southern summer turned the region's grasses brown, Lee's horses often went without suitable forage. In mid-August, Lee noted that he could not advance for "fear of killing our artillery horses," due partially to hot weather, but mainly because the animals had little food. By then, most of the horses survived on an average of one pound of corn per animal per day. No animal got more than five pounds on any given day. Although the concentrated energy in the grain could sustain horses for a limited time, a continuous diet of corn often led to digestive problems such as colic and under certain circumstances might cause an animal to founder. With a near steady supply of better feed from the North's various supply depots, Union animals stayed much healthier even in the hottest weather.[32]

During its invasion of Pennsylvania, the Army of Northern Virginia had precious little time to acquire horseshoes or the material for making them in the field. On the march, Confederates had to abandon many poorly shod horses that could not keep up. Even after returning to Virginia, the cavalry remained in a "shattered state" because of "the want of horseshoes and horseshoe nails." Lee also discovered that the horseshoe shortage put a halt to fully half of his artillery and supply wagons, but he had long since developed strategies for coping with chronic supply problems. When his army lost half its artillery horses at Antietam, Lee had reduced his number of batteries from seventy-five to fifty-four. In January 1863, he trimmed his batteries from six guns to four guns each. The smaller batteries needed fewer horses, but they lost one-third of their firepower. After Gettysburg, Lee had to tighten his artillery belt again because of the lack of forage and horseshoes, writing to President Davis in late August 1863, "I have been obliged to diminish the number of guns in the artillery, and fear I shall have to lose more." The horseshoe shortage continued to plague southern armies well into autumn. On at least one occasion, artillery officers "stripped the shoes and saved the nails from all the dead horses, killing for the purpose all wounded and broken down animals."[33]

Bereft of the horsepower to move his guns and wagons, the normally aggressive Lee assumed a defensive posture in northern Virginia. With little to gain there, President Davis sent part of Lee's army to Georgia in early September to try to reverse southern fortunes on the war's western front. Led by James Longstreet, the detachment contained 15,000 men and essentially no horses. Longstreet's horseless corps had orders to link up with the Army of Tennessee commanded by Gen. Braxton Bragg, who, after retreating from the bone-dry terrain of Perryville, Kentucky, in the fall of 1862, abandoned Middle Tennessee at the turn of the New Year after a bloody battle at Stones River. The victor of that battle, Gen. William Rosecrans, who had replaced Don Carlos Buell in command of the Union Army of the Cumberland in November 1862, did not immediately pursue the Confederates, but he eventually drove Bragg's men into Chattanooga in early July and into north Georgia two months later. Needing greater mobility to confront Rosecrans, Bragg began to outfit his forces with any horses and mules he could find.³⁴

In July 1863, Bragg dispatched a brigade to obtain the necessary animals in Atlanta, which had more horses and mules than usual at that time. In the preceding weeks, local people, in fear of Yankee cavalry raids, had brought their animals into the city for safekeeping. Now the Confederate army effectively cordoned off the entire city. Rebel guards blocked the main roads out of town to keep citizens from sneaking their horses away. Quartermasters took the best animals for cavalry and artillery duty and sent the remaining stock to pull supply wagons. Upset with their army's "high handed and dishonest" tactics, many in the city tried to hide their animals, usually with little success. One Mississippi sergeant reported that his men "got fine carriage horses out of parlors, from sitting rooms, and in one instance from up stairs."³⁵

With temporary equine resources reluctantly supplied by Georgia civilians, Bragg launched an attack against Rosecrans on September 18 in what became known as the Battle of Chickamauga. Only the leading elements of Longstreet's force had arrived, but over the next two days, a brutal fight ensued in the rocky north Georgia woods. On September 20, Longstreet led his troops through a fortuitous gap in the Union defensive line, rolling up the Federal position, and leading to a rare Confederate victory in the western theater, though at a painfully heavy cost. Bragg lost 18,500 of his 65,000 men while Rosecrans lost 16,000 of 60,000. As the defeated Union army staggered back toward Chattanooga, Longstreet urged Bragg to swing around Chattanooga and immediately advance on their supply base of Nashville to reclaim Middle Tennessee for the South. It was a daring plan and one that might have dramatically altered Confederate fortunes in the west, but Bragg could not execute it. Fully one-third of his horses and mules, including many of those recently appropriated in Atlanta, had been killed at Chickamauga. The army simply could not move fast enough, and Longstreet's corps, with plenty of men but no animals, provided little help.³⁶

Instead, Bragg decided to lay siege to Chattanooga, an alternative strategy that had enormous implications for animals in both armies. Bragg placed his troops on the high ridges overlooking the city and cut off its main supply line—the twenty-seven-mile road to Bridgeport, Alabama, which ran along the south bank of the Tennessee River. Longstreet's men took control of the route, forcing the Union army's commissary wagons to use a road sixty miles long that snaked its way across the Cumberland Mountains, quickly gaining and losing more than 5,000 feet in elevation. Natural forage was scarce and resupply often proved impossible. In short order, Union animals began to experience the same kind of deprivation that had become common among the Confederacy's horses and mules. As one Ohio soldier wrote, "It was no unusual sight to see trees as high as animals could reach, barked and eaten as food."[37]

Heavy rains commenced in October, creating mud that reached to the bellies of the malnourished equines. Just to get through the mud, "teamsters harnessed sixteen mules to a wagon and assigned two soldiers to each mule—one to pull on the mule, and one to get behind the mule and push." Southern cavalry constantly raided the shaky supply route. In one memorable attack in the first week of October, Confederate general Joseph Wheeler's cavalry struck a Union wagon train that extended nearly ten miles over the narrow roads. Wheeler killed 1,000 horses and mules and captured nearly as many more (though pursuing Federal cavalry recaptured most of those). In beleaguered Chattanooga, neither man nor beast could find adequate rations. Union officers had to place guards over the animals' feed to prevent starving soldiers from stealing the corn that northern horses now had to eat. By mid-October, one Iowa officer noted, "the mules are dying by the fifties every day for the want of grain."[38]

Thoroughly frustrated with Rosecrans, who acted "confused and stunned, like a duck hit on the head," Lincoln finally ordered Gen. Ulysses S. Grant to assume command of the troops on October 18. By the end of October, Grant had reopened the direct supply line to Bridgeport. For most of the starving animals, the move came too late. By the time the siege ended, one observer noted, "a stinking wall of 10,000 dead horses and mules" ran sixty miles from Chattanooga to Bridgeport.[39]

Events at Chattanooga illustrate a final ecological problem inherent in the use of equine energy to move armies, the same one Lydia Leister faced at Gettysburg. A dead horse became a health hazard. In warm weather, metallic green- or copper-colored blowflies showed up within minutes of the animals' demise, even during battle. Those insects have an uncanny ability to locate the decaying flesh that facilitates their reproduction. In fact, they are so quick to arrive that modern forensic entomologists use their presence to determine time of death. The flies lay eggs that, in as little as twelve hours, hatch into the larvae commonly called maggots. Some estimates suggest that a single dead horse can support up to 400,000 maggots; the voracious larvae can expose portions of the equine skeleton in only five days. The maggots then pupate in the surrounding soil and a new generation of adults emerges six to eight days later. The adults feed on sap or nectar, returning to a carcass to lay another round of eggs.⁴⁰

Blowflies play an essential role in the natural process of decomposition, but because they move rapidly between their breeding grounds and human residences, they can quickly spread bacteria. Given the importance of animals to nineteenth-century life, piles of dung, another powerful blowfly attractant, could be found around every stable and dwelling. The excrement and filth common to military encampments drew blowflies by the tens of thousands. Even early in the war, without the human and animal carnage common in battle, flies swarmed over soldiers' tents and awakened recruits "with tickling sensations about the ears, eyes, mouth, nose, etc., caused by the [insects'] microscopic feet and inquisitive suckers." Under such conditions, blowflies became not just nature's decomposers but also ideal vectors for bacteria that caused dysentery and typhoid. Human corpses left unburied or hastily interred posed the same threat.⁴¹

Nineteenth-century people did not understand the ways the deadly ecological combination of insects, carrion, and unsanitary conditions spread disease. They rightly regarded dead animals as a threat to health, but options for disposing of large animal carcasses were limited. Even today, horse owners have only three main methods for dealing with a dead animal: bury or compost it, incinerate it, or leave it to nature's scavengers and the elements. The last is a practice outlawed in thickly settled regions but still employed in parts of the American West when an animal dies far from home. By the 1860s, officials in urban areas employed all those methods and sometimes tossed carcasses into the ocean or large streams where, theoretically, moving water helped decompose the flesh and dilute its effects on people.⁴²

The most common disposal method employed during the war was abandonment, the same fate that befell most of the animals that perished in the siege of Chattanooga. Armies on the move pushed the carcasses off into the woods or fields away from major thoroughfares so as not to impede foot or wagon traffic. Injured animals unfit for further duty might be dispatched on the spot. During McClellan's retreat during the Seven Days' Battles in late June 1862, a Union cavalry regiment drove 230 injured horses and mules into swampy woods where, according to one observer, "the poor animals [were] slaughtered by lance and pistol" to prevent their capture and use by pursing Rebel soldiers.⁴³

The death of so many horses on major battlefields, however, created a different set of problems. When armies remained encamped near the sites or civilians returned to their battered homes and fields, dead animals became yet another source of contaminated water and its associated diseases. Even interment of the dead animals did not always help. Modern guidelines suggest that a single horse carcass should be placed in a trench seven feet wide and nine feet deep to insure environmentally safe disposal. Few Civil War commanders had the time or inclination to go to such lengths in the wake of battle. When they attempted less sophisticated burials, their men suffered the consequences. Bivouacked on swampy terrain after the Battle of Fair Oaks in June 1862, the Seventh Michigan Infantry dug pits for dead comrades and their mounts in the miry Virginia soil. They soon discovered, however, that "we could not bury them deep enough because we got drinking water at the depth of a spade. Whenever it rained, bodies would rise up, both horses and men. Then we would have to go in and cover them all over again." At Perryville, soldiers desperate to quench their thirst sometimes found streams polluted by carcasses of horses that had been drawn to the water, probably just before they died from wounds suffered in battle. Even one of the horses left on the Leister place near Gettysburg apparently poisoned the spring that provided the family with drinking water. Like those who disposed of dead animals in major cities, certain commanders ordered dead horses thrown into larger waterways. Some of the animals that perished in the siege of Chattanooga probably ended up in the Tennessee River, a stream large enough to serve as a convenient dumping ground.[44]

When dead animals could not be abandoned or thrown into a nearby river, commanders ordered the carcasses to be burned on-site. Groups of soldiers dragged dead horses and mules into piles, heaped brush and limbs on the animals or soaked them in lamp oil, and put the torch to the heap. The smell of burning hide, hair, and flesh usually proved no more pleasant than the odor of the maggot-infested carcasses. After the Battle of Cedar Mountain, Virginia, in August 1862, a Union general ordered all the dead horses burned, but bystanders complained that the smell of "roasted carrion instead of raw" now permeated everything for miles around. Less than a week after the Battle of Antietam, a Union surgeon recorded that wounded soldiers had to be quickly removed from hospitals near the battlefield because "the odor" of hundreds of burning horses was too "horrible" to tolerate. The horses on the Leister farm became part of an estimated 5 million pounds of horseflesh burned after the fighting ended at Gettysburg. Ridding the area of dead horses proved such loathsome duty that army officials often used it as punishment, sentencing those caught pilfering items from the battlefield to the disposal details.[45]

At major depots, where animals frequently died of disease, command-

Burial parties collecting corpses and burning dead horses after the Battle of Fair Oaks, June 3, 1862, near Richmond, Virginia. Incineration proved more efficient and hygienic than burying or abandoning the animal carcasses.
Courtesy Library of Congress.

ers did order horses buried, perhaps to prevent the spread of glanders and farcy. At Giesboro, where as many as 300 animals died every day, a standing detail of twenty-six soldiers took care of the burials. In addition, the Union contracted with civilian firms that processed the carcasses to extract collagen, a key ingredient in various glues and adhesives. Such businesses also boiled horse skeletons to remove the flesh and then ground the bones for use as fertilizer. Horses still had value, even in death, as the sale of carcasses netted the government about $60,000 between 1863 and 1864. Apparently, agents or factors from such firms showed up at major battlefields, including Gettysburg, to gather what remained of the charred horse and mule carcasses. Lydia Leister, who had suffered much due to the seventeen dead horses left in her yard, eventually sold off their skeletons to one such collector. At the going rate of fifty cents per hundred pounds, the transaction netted her about $3.75, suggesting that she sold about 750 pounds of bones. She and others like her extracted the last bits of remaining energy from the horses that had provided motive power for the war's largest battle.[46]

Equines could transport artillery, supply wagons, and cavalry, but foot soldiers required nutritional energy to propel them into battle. Much of the animal protein that kept the infantry moving came from pigs, usually in the form of salt pork and bacon. When it comes to converting nature's stored energy matter into meat suitable for human consumption, hogs are among the most efficient creatures in the animal kingdom. They have simple digestive systems with large stomachs capable of processing almost anything, making them true omnivores. Recent studies of feral hogs suggest that, left to their own devices, swine routinely consume about 3 to 5 percent of their body weight in food every day. While vegetation makes up the bulk of that diet, hogs routinely eat earthworms, various insects, arachnids, fish, amphibians, and mammals, both dead and live. In a domestic setting, pigs also favor corn, a crop Americans produced in abundance during peacetime. Swine waste no time in transforming such matter into human comestibles — the animals convert their food into meat more than three times faster than cattle.[47]

Pigs reproduce more rapidly than bovines, too. Typically, a cow requires nine months to give birth to a single calf. The gestation period for hogs is less than half that and sows average five to ten piglets in every litter. When Hernando de Soto began his explorations of the American Southeast in 1539, he brought along thirteen pigs to serve as a mobile meat locker. Within just a few years hundreds of the animals roamed the southern countryside. Wherever Europeans introduced them, hog populations exploded, providing future generations of Americans with a reliable source of protein. Compared to beef, pork proved easier to preserve by salting or smoking, making it an ideal food for troops on the move. Both armies adopted it as a standard ration.[48]

In the South, most livestock owners practiced open range management. They fenced their crops instead of their animals, letting swine fend for themselves in woods, swamps, or unimproved fields for most of the year. After harvest, the animals might be directed into cornfields to clean up remaining stalks and other refuse. Though southern hogs sometimes fattened on corn a few weeks before slaughter, much of their nourishment came from wild forage. As a result, they tended to be small and lean, averaging less than 150 pounds, prompting one observer to describe them as "long lank, bony, snake-headed hairy wild beasts."[49]

Those small pigs and their stringy meat proved especially important to yeoman farmers and the southern poor. Laws that required fencing of crops and free access to the woodlands had been on the books in Virginia since the 1600s, ensuring that even those with only a few head of hogs or other livestock could provide meat for their families. Throughout the antebellum years, such folk had successfully resisted efforts by the upper classes to reform fencing laws and restrict such freedoms. As historian Jack Temple Kirby explains it, "The southern countryside was truly distinctive and, ironically, considering the prevalence of plantation slavery, a 'democratic' countryside where even poor men (white, mostly) could feed their families and, as drovers and sellers of surplus beef and pork, participate in markets."[50]

All classes of southerners slaughtered thousands of hogs every winter and utilized "everything but the squeal." What could not be eaten fresh was preserved—not just hams, shoulders, tenderloins, ribs, and bacon but the head, snout, feet, ankles, backbone, brains, heart, lungs, liver, kidneys, tongue, ears, tail, intestines ("chitterlings" or "chitlins"), and the guts, used as casing for sausage. Farmers rendered all excess fat into lard, the preferred southern cooking aid, used to fry anything edible. As Dr. John S. Wilson of Columbus, Georgia, put it, "Hogs lard is the very oil that moves the machinery of life." By 1860, southerners raised two-thirds of the nation's hogs. Seven southern states had over 1 million hogs each. Wilson called the region "the great Hog-eating Confederacy, or Republic of Porkdom."[51]

From mid-October to mid-December, drovers moved hogs out of the Deep South to holding pens in Knoxville, Tennessee, a city that soon emerged as a central distribution point for animals marketed in the Carolinas and Virginia. Each year more than 100,000 hogs found their way to Tennessee and points east. A booming hog population and a thriving regional trade did not, however, provide the South with an economic advantage going into the war. Cultural stereotypes notwithstanding, northerners ate lots of pork, too. And as with other facets of food production, Yankee methods for turning pigs into palatable protein proved better organized and more efficient than those employed in the South.[52]

Aided by fencing laws that kept livestock confined, northern husbandry stressed penning hogs and fattening them on corn in preparation for sale to major meatpacking operations, first in Cincinnati—nicknamed "Porkopolis" for its crucial role in the trade—and later in Chicago. Hogs processed in those cities averaged more than 200 pounds each by 1860, meaning that each animal had roughly 25 percent more usable meat than their leaner and meaner southern counterparts. More efficient slaughtering, packing, and

Cartoon showing cattle and hogs, along with the appropriate condiments, marching into battle with Union soldiers. Such animals provided much-needed protein for men in the ranks. From *Harper's Weekly*, June 1, 1861.

distribution methods also boosted production. In the years leading up to the war, the volume of pork shipped from the Midwest far outstripped that traded in regional centers such as Knoxville. By 1860, of the 3 million hogs commercially slaughtered and packed for market in the United States, only 20,000 came from the Confederate states. That same year, the South, with a seemingly insatiable appetite for pork, actually imported 1.2 million hogs from the Midwest.[53]

The first shots on Fort Sumter shut down that crucial supply of imported pork. Southern leaders were quick to recognize the perils of the looming meat shortage. In 1861, Confederate commissary general Lucius Northrop warned, "The real evil is ahead. There are not enough hogs in the Confederacy sufficient for the Army and the larger force of plantation negroes." He estimated that the southern army alone might require half a million hogs per year just to keep soldiers in the field. He doubted the region could produce more than one-third of that number, a gloomy calculation that did not portend well for civilians accustomed to consuming pork at almost every meal. During the first two years of war, Union advances in the western theater turned Northrop's grim prophecy into reality.[54]

By the summer of 1862, northern armies controlled all of Kentucky and two-thirds of Tennessee, two states that had once provided large numbers of hogs traded in the South. Control of central Tennessee proved especially costly as an important Confederate meatpacking plant at Nashville fell into Union hands. A year later, northern troops captured Chattanooga and Knoxville, depriving the Confederacy of hogs shipped from east Tennessee. In the spring of 1863, repeated Union raids in the Tennessee River valley brought the hard hand of war to rich farmlands in northern Alabama. Making "war upon the land," as historian Lisa Brady argues, became central to Union military operations, and northern generals increasingly began to attack the enemy's ecological base. Yankee strategists understood that hogs, with their ability to turn almost anything in nature into meat, occupied an important niche in the southern food system. As a result, those who wreaked havoc on the southern landscape frequently targeted pigs. In May 1863, Union general Grenville Dodge happily reported that he had either captured or driven away 1,000 hogs and destroyed half a million pounds of bacon, "render[ing] useless for this year the garden spot of Alabama."[55]

Mississippi fared even worse in 1863. The Union navy interdicted trade on the Mississippi River in February, and a single gunboat captured 110,000 pounds of pork and 500 hogs destined for the Confederate garrison of Port Hudson, Louisiana, 150 miles south of Vicksburg. Ulysses Grant's soldiers slaughtered thousands of Mississippi hogs to feed themselves during the Vicksburg campaign. On June 5, a southern officer wrote from the Yazoo River region northeast of Vicksburg that Union troops took "all provisions from the citizens indiscriminately killing cattle, hogs, chickens, and so on, with the evident intention of destroying the subsistence and forage which could be of service to our army." The policy continued even after the Confederates surrendered Vicksburg on July 4. Ten days later, Gen. William T. Sherman wrote from near Jackson, "We are absolutely stripping the country of corn, cattle, hogs, sheep, poultry, everything." He admitted that future hardship for local residents might be "terrible to contemplate, but it is the scourge of war." With the trans-Mississippi region effectively severed from the Confederacy in the summer of 1863, southern soldiers and civilians lost yet another valuable source of the protein that sustained them.[56]

Well before the fall of Vicksburg, the escalating shortage of hogs helped dictate Confederate strategy. In mid-February 1863, southern secretary of war James Seddon urged the military commander of east Tennessee to launch a large cavalry raid into Union-occupied Kentucky to "send back supplies of bacon and salt meat, and drive out large droves of hogs and cattle" to southwestern Virginia, an area "now almost destitute of stock." Cavalry troops embarked on the raid but "found it impracticable to obtain stock or hogs." As early as February 23, 1863, John J. Walker, the southern chief of subsistence, believed that if government reserves of salted meat could not be replenished, the army might run out of that essential ration by June 1. Without meat, especially the salt pork that kept the troops moving, Walker now faced "the terrifying thought of a noble and conquering army demoralized, and possibly disbanded, for want of food," a situation he thought "fearfully near." The only way to prevent such a calamity, he argued, was either for Confederate armies to drive the Union out of Tennessee and Kentucky or, failing that, for the South to begin direct trade with the enemy for meat.[57]

Trading with the Union had been floated in Confederate circles four months earlier when a Memphis merchant proposed exchanging cotton for "ten thousand hogsheads of bacon certainly, and probably twenty thousand hogsheads more." Commissary General Northrop and then–secretary of war George Randolph supported the plan and thought they had President Davis's approval. When Davis changed his mind and forbade the trade, Randolph resigned in frustration. However, by the following spring, civilians and military authorities began engaging in an illicit meat trade on their own, forcing Confederate authorities to accede. In March 1863, Gen. Edmund Kirby Smith began trading cotton for meat at the mouth of the Rio Grande, and Davis accepted the practice, albeit with some restrictions. The trade of fiber for protein largely sustained the South's western armies, and Memphis became such a hub of this commerce that one wag dubbed it "a greater outfitting point for Confederate armies than Nassau" (the Bahamas port frequented by Confederate blockade runners). Over eight months in 1864, nearly $12 million of supplies moved from Memphis into the Confederacy before President Lincoln shut down the trade.[58]

The South also sought hog meat overseas. Confederate commissary agents purchased nearly 3 million pounds of bacon from England in 1863, but by the end of the year none of it had arrived due to myriad transport problems. By November 1863, the Confederate Commissary Bureau had only 2 million pounds of bacon on hand, roughly one-third the amount available just one year earlier. Meanwhile some civilians in Richmond stole salt pork from their neighbors and scores of soldiers wrote to the secretary of war threatening desertion unless the government provided their families with adequate food. Confederate officials could provide little relief, writing in December 1863, "The last pound [of meat had] been forwarded to General Lee's army," leaving Virginians with "no 'reserve depot' to draw upon."[59]

In their wide-ranging search for pork, commissary agents accidentally unleashed another scourge, one that posed a far greater threat to southern hogs than even the most ruthless Yankee raiders. Just as the mustering of horses spawned the glanders epizootic, the transport of pigs from distant locales aided in the transmission of hog cholera, a devastating porcine disease. The virus responsible for hog cholera probably came to North America with pigs imported by early European settlers. However, the organism seems to have remained a minor threat until the development of larger midwestern meatpacking operations in the 1830s. Confinement of so many hogs in a relatively small space probably allowed for mutation and development of more powerful strains of the virus. The telltale signs of infection included fever, vomiting, and diarrhea. Some animals died of dehydration. Others seemed to recover, only to suffer relapse and sudden death. Isolated reports of hog cholera surfaced in southern states in the 1830s and 1840s, but by 1858 an epidemic in Tennessee dropped the state's hog population by about a third.[60]

Once established south of the Ohio River, hog cholera spread rapidly. It not only passed from animal to animal through contact but also lingered in processed pork. Pigs that ate scraps discarded by humans might become infected. Skeletal remains from hogs that perished from the disease could also pose a threat. Worse, animals that showed no symptoms sometimes carried and spread the virus. As with glanders, the war became a nearly ideal vector for the infection. In the early stages of the conflict, officials sent hogs from across the South to holding pens and processing stations in and around Richmond in an attempt to provide some fresh food for troops at the front. Hogsheads of salt pork from distant locales made their way by wagon and rail across the South. As a result, the virus had a nearly inexhaustible supply of hosts, at least until military and civilian demand, as well the ravages of the disease itself, began to put a dent in the swine population. Hundreds of thousands of hogs died every year from the disease, and it largely annihilated the meat supply in Arkansas in 1862. The disease left a lasting legacy in the Confederate states. Not only did the Republic of Porkdom never recover its numbers, but, as Erin Stewart Mauldin argues, efforts to contain the disease also wiped out stock-raising opportunities for poorer southerners. In the decades after the war, southern state governments initiated fence laws that required the penning of livestock, removing a traditional southern subsistence practice that had existed for centuries.[61]

In the never-ending effort to keep their armies fed, the Union and Confederacy also made use of beef. Southerners raised cattle much as they did hogs, allowing the animals to roam the open range and feed on whatever nature provided. Cows are herbivores and ruminants. The plants they consume quickly pass into a highly specialized digestive chamber, called a rumen, where various bacteria begin to break down and ferment plant matter. The animals then regurgitate the mixture, known as a cud, and chew it more thoroughly before it passes into the stomach and small intestine for further digestion. Like deer and other wild ruminants, cattle can survive by grazing on a wide variety of trees and plants native to the South. Bovines found sufficient food in hardwood forests, on natural savannas, and especially in canebrakes where the animals grazed on tender shoots of indigenous bamboo. Though not nearly as prolific as hogs, cattle had little trouble adapting to the southern environment. Also valued for their hides, cattle became a vital source of food and income, especially for those who lacked the resources to invest heavily in land and slaves.[62]

In the first decades of the nineteenth century, southern herders began to take advantage of forests and grasslands on the western frontiers of Alabama and Mississippi. In time, encroachment of cotton planters pushed cattlemen into the piney woods and less fertile areas of those states. By 1860, many herdsmen had moved west of the Mississippi River, but cattle remained essential to the southern economy. On the eve of the Civil War, southern cattlemen owned some 9.25 million animals. At least a third of those and perhaps as many as half were semiferal Texas longhorns. Visitors to the antebellum South not only took note of the large herds but also observed that the region's cows—like its hogs—tended to be lean and bony and sometimes appeared to show signs of malnutrition. Even when southerners rounded up and penned cattle to fatten them for market, the animals usually ate only corn blades, "cheap roughage" that "came straight from the cornfield with no extra effort such as seeding, cultivating, or mowing." As one writer has noted, "While almost certainly preferable to the cow itself," southern husbandry "hardly resulted in tender cuts of beef come slaughter time." In addition to producing poor meat, the lack of nutrient-rich grasses also led southern cows to produce a limited quantity of inferior-quality milk. Given the labor-saving ease of raising cattle on the open range, it was a tradeoff that most southern cattlemen were willing to accept.[63]

Left to fend for themselves in forests and fields, southern bovines faced a number of potential hazards. Some animals wandered away, became trapped in swamps, drowned in flooded rivers, or simply went feral. In 1860, the greatest threat to cow health and vitality in the region probably came from unseen organisms transmitted by one of the South's most loathsome parasites. As cattle foraged in southern forests and grasslands, the animals inevitably became infested with a variety of blood-sucking ticks. Some of those ticks, members of the genus that transmit Lyme disease, likely carried protozoa that affected cattle much like malarial parasites affect people. The most serious group of microorganisms, protozoa known collectively as

Babesia, lived in tick saliva. Transmitted to cattle via a tick bite, the microorganisms attacked bovine red blood cells, resulting in "high fever, anemia, listlessness, trembling, delirium and grinding of teeth." The infection, today called *Babesiosis*, can kill an otherwise healthy cow in fewer than ten days. In a herd of previously unexposed animals, overall mortality can run as high as 90 percent. Another less lethal group of tick-borne parasites, *Anaplasma marginale*, might also have been present in the antebellum South. They cause a condition known as *Anaplasmosis*. It produces similar but less severe symptoms and is usually lethal only among older cattle. However, as with malaria in humans, *Anaplasma marginale* lingers in the bovine system, leading to chronic anemia and listlessness. It is unclear how many southern cattle suffered from these diseases during the war, but they were a constant concern for authorities.[64]

While southern cattle settled in on the open range, their northern counterparts adapted to more temperate conditions. Thousands of years earlier, advancing glaciers had left the soils of New England thin and rocky. Those who sought to make a living off what remained turned to grass and large animals. Early European settlers gave over much land to permanent meadows, made possible by a generally cool, wet climate that facilitated the growth of various grasses, including those transported from Europe in the food and manure of imported livestock. Once established on the land, cattle and sheep turned grass into manure that could be collected in barns and spread on fields to restore and maintain fertility. Wheat, rye, oats, and barley — grasses that had been adapted to human use — grew in fertilized fields. In time, northern farmers began to pen livestock on pasturelands, where the animals grazed on a variety of plants and grasses, including alfalfa and timothy. The system required much year-round labor, but it proved durable and, in a practical sense, recognized the ecological relationships that might help sustain northern agriculture.[65]

It also produced well-fed bovines in abundance. Cattle might grow and breed on the South's open range, but the animals are far better equipped to eat grass. Unlike deer and other wild ruminants, cows lack upper teeth that aid in stripping leaves from trees. Cattle also have thinner saliva, larger rumens, and generally more specialized digestive tracts, evolutionary adaptations that make it easier for the animals to break down cellulose in grass. Some modern estimates suggest that it takes fifteen to thirty acres of woodland to provide a single cow with food value equal to a single acre of improved pasture. Like cattle in the South, northern bovines might come into contact with several varieties of ticks and parasites. But thanks in part to the

Beef for the Army—on the March, etching by Edwin Forbes, 1864.
Courtesy Library of Congress.

climate, the protozoa responsible for *Babesiosis* and *Anaplasma* could not survive in more northerly latitudes, thereby affording the Union yet another ecological advantage in animal husbandry.[66]

Beef, dairy products, and grain not only kept northerners fed but also found markets in the region's emerging cities. Even so, many in the North preferred mobility to long-term management. As population pressure increased, northern farmers and newly arriving immigrants increasingly sought subsistence and profit in the richer soils of the Midwest. When the war began, some 13.5 million cattle grazed on Union lands, roughly one and a half times as many as foraged on the South's open range. By all accounts, northern cattle were fatter, too. Animals sent to market from the Ohio Valley typically weighed between 1,000 and 1,500 pounds, while southern cattle rarely reached 800 pounds.[67]

Southerners typically valued cattle more for milk, butter, and leather than for beef, largely because of the difficulties inherent in preserving the meat. A butchered cow provided far too much food to consume fresh, and when it came to preserved meats, southerners much preferred salt pork to salted beef. Before the war, a typical southerner ate roughly 25 pounds of beef annually, compared to 150 pounds of pork. Even so, when Confederate commissary officers realized how few hogs the South could provide for its armies, beef became a much more attractive source of animal protein. Moreover, cattle were far easier to drive overland than hogs, meaning they could provide fresh meat on the hoof, while hogs usually had to be slaughtered locally and the salted meat shipped to the front in barrels.[68]

Texas had an immense number of longhorns, but it was a long way from the Virginia front. Cattle drives took weeks, and the rickety southern railway system proved equally unwieldy in transporting beef to eastern armies. After the lean winter and spring of 1863, Lee's men moved into Pennsylvania in June, desperate to replenish their commissary stocks. As Confederate troops marched north, they collected 1,000 head of cattle at Hagerstown, Maryland, and rounded up another 3,000 head at Chambersburg, Pennsylvania. Federal forces in pursuit of Lee's retreating army after Gettysburg found the remains of hundreds of slaughtered cattle at southern camps along the way, indicating that the hungry Rebels frequently ate fresh beef as they moved through the region.[69]

Meanwhile, Union military successes in Mississippi that summer played a decisive role in the worsening Confederate food crisis. After taking Vicksburg, some Union forces moved south to Natchez, Mississippi, seizing 5,000 head of cattle that had recently crossed the river to be sent to southern armies. Union quartermaster general Montgomery Meigs confidently proclaimed that the capture of Vicksburg, which cut off nearly all supplies coming from Texas, "must have a most important bearing on the war." Indeed it did. Denied the seemingly endless supply of Texas beef—and with supplies in other southern states either compromised by Union depredations or depleted by constant commissary requisitions—southern armies east of the Mississippi turned to Florida as their last untapped source. Throughout the autumn of 1863, army commissary agents repeatedly begged Florida subsistence officers to send them cattle to provide their soldiers' daily meat rations, since bacon was in such scarce supply. One officer admitted in November 1863, "Now, two large armies look almost solely to Florida to supply one entire article of subsistence." Another commissary officer suggested that Florida cattlemen ship 1,000 head of cattle a week to meet the armies' demands. Confederate armies quickly exhausted that supply, as well as virtually any cattle that could survive the overland journey to processing plants in Georgia and the Carolinas.[70]

Hunger was not the only problem that resulted from the South's dearth of cattle. Cowhides provided raw material for leather, a vital war matériel used in saddles, harnesses, bridles, reins, and especially shoes. While the North produced ample amounts of leather throughout the war, the South started experiencing shortages as early as 1861, largely because it had imported the bulk of its leather from the North. Trade with suppliers in Texas, Mexico, and South America dried up thanks to the Union blockade. Additionally, the enlistment of so many tanners into the army in 1861 significantly curtailed the processing of leather. Eventually, the Confederate government exempted tanners from military service, but shortages of raw materials inhibited adequate production. The primary government leather shops, located in Montgomery, Alabama, could not acquire enough hides to meet the enormous army demands. Necessity prompted some manufacturers to make ersatz bridles and saddle skirts from multiple layers of cotton cloth stitched together.[71]

On the march to Gettysburg, Lee's quartermasters purchased 5,000 pounds of leather from merchants in Williamsport, Maryland. The supply barely made a dent in the army's needs, especially its crippling shortage of shoes. In January 1864, Lee sent his army quartermaster to purchase or impress leather in the Shenandoah Valley and requisitioned at least 37,500 pounds of leather from the quartermaster bureau. The general desired to set the shoemakers in his army to work making footwear for the soldiers, because he believed that shoe factories in Richmond and Columbus, Georgia, made inferior products that "would not stand a week's march in mud and water." Despite their best efforts, quartermasters could never meet the army's requirements.[72]

While Union military encroachments and extensive consumption of beef led to a substantial loss of southern cattle during the war, adverse weather also caused devastation in the western parts of the country. In the brutal winter of 1863–64, arctic blasts swept through Texas, killing anywhere from 50 to 90 percent of the cattle on some farms. But the unusually severe Texas winter paled in comparison to the drought that plagued California beginning in the war's second summer. After the unprecedented storms and flooding in early 1862, the California landscape baked beneath a scorching sun. The drought lasted three long years. Altogether, half a million California cattle perished during the war.[73]

The death of millions of animals was a terrible consequence of the conflict that would linger for decades, but the death and maiming of hundreds of thousands of soldiers would resonate far more powerfully with the generations that followed the war. Enfeebled by wounds, disease, or exposure, addicted to painkillers, and haunted by the stress of combat, damaged soldiers emerged from the war in staggering numbers. Nearly three-quarters of a million people perished during the four years of conflict. No seasons would prove more terrible for death and disability than the spring and summer of 1864.

Five

DEATH AND DISABILITY

SPRING 1864–FALL 1864

In the early morning of May 12, 1864, the soldiers in Gen. Nathaniel H. Harris's Mississippi Brigade awoke to rain and the sounds of battle. Three corps of the Army of the Potomac had launched a dawn attack against fortified trenches and earthworks defended by Confederate troops at Spotsylvania, Virginia. Known as the "Mule Shoe," the defensive position was a "salient"— a distinctive part of the earthworks that extended a mile or so out from the rest of the Rebel line—in a curve like a mule shoe. The initial Union attack breached the fortifications and captured 3,000 soldiers, but the stunned defenders rallied to check the invaders. Robert E. Lee hurried reinforcements, including Harris's brigade, forward to hold the line. As they made their way to the front, a member of the famed Stonewall brigade shouted to the Mississippians, "Boys, you are going to catch hell today." That prediction came true with a vengeance.[1]

Harris's brigade rushed into the breach near the northwest corner of the salient, thereafter fittingly known as "the Bloody Angle." As one Mississippian recalled, "The enemy seem[ed] to have concentrated their whole urging of war at this point. Shells of every kind and shape . . . [and] a forest of muskets played with awful fury over the ground itself." The 800 tired and malnourished southerners resolutely marched into the fusillade, prompting a Confederate general to remark, "It was the bravest deed I have ever seen performed." Corps commander Richard Ewell similarly expressed his "intense admiration for men who could advance so calmly to what seemed and proved almost certain death." As Harris later reported, for twenty hours "my men were exposed to a constant and murderous musketry fire."[2]

Unrelenting rain turned the battlefield into a quagmire. Heavy fog and dense, sulfurous smoke blanketed the landscape, limiting visibility to a few dozen yards. The fighting often occurred in such close quarters that soldiers employed bayonets, knives, and axes in hand-to-hand combat. One Confederate witnessed a "tall, brawny fellow" throw away his musket and pluck a hatchet from the ground: "As a Federal comes at him with a bayonet, he pushed it aside with his left hand, while with the hatchet in his right he brains his opponent." Bayonets pierced vital organs; musket butts crushed skulls; minié balls tore away flesh, shattered bones, and sprayed fountains of blood and brain matter over the combatants. In the frenzy, some soldiers jumped on top of the parapet to fire pointblank at their opponents and hurl their bayoneted rifles like spears into the massed enemy before being shot down. Wounded men fell into water-filled trenches that soon ran red with blood. Some of the injured slowly drowned, pressed beneath the surface as comrades clamored over their mangled bodies. The sheer volume of lead tearing through the air cut down an oak tree nearly twenty-two inches in diameter just to the rear of the trench line. One soldier wrote that as the "cold drenching rain continued" throughout the battle, "the flashing lightning, the bursting of shells, the tremendous and incessant roar of small arms, and the yell of the soldiers presented a scene indescribable in its terrible horror." When the fighting finally ended about 3:30 a.m. on May 13, Harris's brigade had lost half of its number in the struggle.[3]

At first light, one observer noted that most of the dead had been "chopped into hash by the bullets, and appear[ed] more like piles of jelly than the distinguishable forms of human life." Another recalled, "The field presented one vast Golgotha in immensity of the number of the dead." A general noticed that "fallen men's flesh was torn from the bones and the bones shattered." Ulysses S. Grant's aide, Horace Porter, recalled that the dead soldiers "exhibit[ed] every ghastly phase of mutilation." But he was particularly horrified by the fate of those not killed outright: "Below the mass of corpses, the convulsive twitching of limbs and the writhing of bodies showed that there were wounded men still alive and struggling to extricate themselves from the horrid entombment."[4]

On the foggy morning of May 13, survivors of the previous day's savage fight — which cost nearly 9,000 Union and 8,000 Confederate casualties — struggled to process the sights and smells that assaulted their senses. The stench of bloated and putrefying bodies was overpowering. One southern soldier recorded, "If a man wants to see hell upon earth, let him come

Painting of the Battle of Spotsylvania showing the fighting at the Bloody Angle on May 12, 1864. Courtesy Library of Congress.

and look into this black, bloody . . . horrid corruption of rotting corpses, that fill the air with their intolerable stench." In one 200-square-foot area of the trenches, Union soldiers discovered 150 dead Confederates piled several layers deep. Soldiers unceremoniously left the dead in their trenches and covered them with a thick layer of dirt from the parapet. "The unfortunate victims," one soldier explained, "had unwittingly dug their own graves."[5]

Even after three years of war, the most hardened soldiers found the sheer magnitude of death difficult to comprehend. Vermont private Wilbur Fisk spoke for many when he noted that, "though I have seen horrid scenes since this war commenced, I never saw anything half so bad as that." Returning home when hostilities ended, veterans found the death and destruction impossible to describe. As one Maine soldier explained, "I never expect to be fully believed when I tell what I saw of the horrors of Spottsylvania [sic], because I should be loth to believe it myself were the case reversed." For any soldier who still maintained visions of fighting as an exciting and glorious adventure, Spotsylvania put an end to such naiveté. Fisk thought of his own mortality while viewing "such a terrible, sickening sight" and admitted,

"I have sometimes hoped, that if I must die while I am a soldier, I should prefer to die on the battle-field, but after looking at such a scene, one cannot help turning away and saying, Any death but that."[6]

By the spring of 1864, men who enlisted for glory and adventure had become keenly aware that they faced the likelihood of a gruesome, agonizing, and perhaps anonymous end. Approximately 750,000 soldiers died in the Civil War, a death toll higher than the *combined* loss of life in all other U.S. wars from the Revolution to the present day. In terms of percentage of total population, the death rate was eight times higher than that incurred during World War II. Even in a society more accustomed to death than ours, the scale of it was unimaginable. Some 2.4 percent of the American population perished in just four years. That would be the equivalent of more than 7.4 million Americans dying in a four-year war today. Additionally, death in war often came suddenly. Young men in the prime of life died in battle, in their tents, in hospitals, and sometimes after returning home. An additional 50,000 deaths among civilians — or nearly 500,000 in today's population — meant that, as historian Drew Gilpin Faust writes, "Death's threat, its proximity, and its actuality became the most widely shared" of Civil War experiences. That new familiarity with death transformed the nation's "most fundamental assumptions" about mourning, burials, and the proper relationship with God.[7]

War also forced both combatants and civilians to confront death as an ecological event. When a soldier breathed his last, his flesh and bones became subject to a variety of new biological processes that involved a vast array of other organisms. Enzymes inside the dead man's body immediately began to consume dead cells, initiating decomposition. Bacteria, too, fed on the cells, especially in the gastrointestinal tract, where organs and soft tissue broke down quickly. In one of the more bizarre facets of initial decomposition, the pancreas digests itself. Within about two days, bodies began to bloat as the dead matter released hydrogen sulfide, ammonia, methane, and a host of other noxious gasses. Various fluids tinged with green and purple, also by-products of bacterial action, escaped through the mouth and nostrils. As many as 400 organic compounds created during the first seventy-two hours of decomposition combined to produce an odor all too familiar to Civil War soldiers: the sickly sweet aroma of putrefaction, the scent of death. If bodies remained exposed to air and the elements, the stench endured for a month or more.[8]

The smell proved difficult to describe. Sometimes it seemed to "come up as if in waves," stopping one's breath "in the throat" and causing a "sense of suffocation." It might also strike suddenly, knocking experienced soldiers to their knees and leaving them with their "mouths close to the ground vomiting profusely." Some of those sent to retrieve the dead used various scented oils and perfumes to mask the odor; others crammed fresh leaves into their nostrils. Nothing, it seemed, allowed for a fresh breath. Only whiskey provided a measure of relief. As more than one soldier noted, it was much easier to approach the dead while "staggering drunk."[9]

Though it seems to defy logic, an individual rotting corpse probably posed comparatively little threat to the health of the living. Those who moved the bodies had a slight risk of contracting tuberculosis, hepatitis, and bloodborne maladies, but only if the dead man suffered from those contagions and only if the corpse was fresh; organisms that cause such diseases cannot survive for more than a day or two after death. Like dead horses and mules, decaying bodies simply contributed to the general filth and unsanitary conditions created by armies on the move. Given that corpses can release feces and various liquids capable of contaminating nearby water supplies, the dead did provide an additional source of typhoid and various diarrheal ailments, but those illnesses were already so well established in the camps that the overall effect of decaying bodies might well have been minimal.[10]

Nineteenth-century Americans, however, did not know that. In the 1860s, most American physicians and army surgeons subscribed to some variation of the miasmatic theory of disease. In general terms, the theory held that sickness resulted from breathing in toxic matter suspended in the air. Such noxious fumes might come from swampy ground or from overall filth and poor sanitary practices. Experts believed that whatever its source, a potentially lethal miasma could be identified by its putrid odor. It was natural to assume that a decomposing body, with its unmistakable stink, released rotten matter into the atmosphere. As a result, whenever people died en masse, removing the reeking bodies from contact with air became a top priority.[11]

Those charged with cleaning up after major engagements had other reasons to move quickly. Long before humans caught the scent of death, the compounds responsible for the stink (perhaps sulphur- or ammonia-based) drew other living things to the decomposing bodies. Viewed from an ecological perspective, the animals simply took advantage of a sudden nutritional windfall. A fundamental principle of ecology holds that whenever new biomass appears, other organisms inevitably show up to consume and utilize it. Hogs were perhaps the most visible consumers, feasting on dead bodies after battles, but other fauna joined the buffet. In much the same way that they colonized the carcasses of horses and mules, various species of flies appeared within minutes of a soldier's death. The insects laid eggs or deposited young larvae directly onto the cadaver, often choosing a wound or other opening in the skin. In warm weather, a body left exposed for three days became home to hundreds of maggots that, over the next two weeks, converted human biomass into several new generations of flies. Those insects had the potential to spread disease, especially if they visited manure piles. But provided with an abundance of dead flesh, flies often spent their entire lives on and around the corpse. Besides, flies already infested the camps, so much so that any uptick in their population from the new food supply might well have gone unnoticed. In time, a variety of ants, mites, beetles, wasps, and spiders preyed on the maggots and helped consume the remains. Two to three weeks after death, such activity left behind bones, hair, bits of tissue, and small pools of smelly liquid that contemporary observers sometimes referred to as "carrion grease." In the absence of vultures or other scavengers, it took perhaps another month or two until the once fleshy cadaver became a skeleton. A fallen soldier might be dead, but once his rotten corpse became part of the natural world, it sustained a surprising array of life.[12]

Decaying fly-infested bodies nearly always provoked comment even in the heat of battle. At Fair Oaks during the Peninsula Campaign, members of the Eighth Illinois Cavalry came upon "hundreds of the rebel dead" that "lay rotting above ground, which was literally covered with maggots crawling in all directions for more than half a mile." Union officers assigned Confederate prisoners of war to dig graves, but the reluctant captives "merely threw a few shovelfuls of dirt over each corpse where it lay." Overnight downpours exposed the bodies, leaving them "partly uncovered, with the flesh falling from the bones and [again] crawling with maggots." The stench came back, too, "filling the hot air with poisonous exhalations."[13]

Dead horses and mules could be incinerated along with their attendant insect life. But for a variety of cultural and religious reasons, even the most maggot-infested human corpse usually required a grave. As Drew Gilpin Faust writes, a "decent burial" helped preserve a corpse's identity and "the promise of eternal life." Among nineteenth-century Christians, such notions served to differentiate human remains from those of animals "who possess[ed] neither consciousness of death" nor the prospect "of either physical or spiritual immortality." On the battlefield, constant exposure to the ecological consequences of death often blurred that distinction, especially if a corpse wore a different uniform. Walking the ground two days after the Battle of South Mountain in Maryland, Union surgeon Daniel Holt found that "none but confederates were left upon the ground—our dead and wounded having been carried off during the evening and night of the conflict." Looking on the "putrifying [sic] bodies of the [southern] dead," with "their protruding bowels, glassy eyes, open mouths, ejecting blood and gases," Holt felt "as little unconcerned as though they were two hundred pigs." He attributed his "harder hearted" nature to a new familiarity with such scenes and his growing disgust with the Confederacy for starting the war.[14]

By the time Holt witnessed the carnage at South Mountain in September 1862, disposal of human remains had already become a haphazard process. Fallen officers might get special treatment, but funerals for enlisted men usually consisted of brief prayers and removal to the nearest ditch or other natural edifice. After Gettysburg, where the July heat hastened decomposition, Yankee and Rebel corpses alike initially went into hastily dug trenches "seven or eight feet wide and about three feet deep." In the "valley between the Round Tops," where soils were thin and stones plentiful, burial details simply piled three or four dead Confederate soldiers together and covered them with rocks and brush, a tactic that "only partially concealed the decomposing corpses beneath."[15]

Such makeshift tombs did little to slow nature's processes or reassure those who associated proper interment with a Christian afterlife. At Gettysburg, the prospect of swine consuming the dead indirectly contributed to the construction of more permanent graves and, coincidentally, one of the war's most famous speeches. Three weeks after the battle a local lawyer wrote the governor of Pennsylvania asking for a new cemetery where the Union dead might be properly interred. As part of his appeal, he noted "several places where the hogs were actually rooting out the bodies and devouring them." His argument helped persuade officials to create what became Gettysburg National Cemetery. At its dedication in November 1863, President Abraham Lincoln delivered a two-minute oration now known as the Gettysburg Address. By mid-March 1864, workers exhumed and reburied 3,354 Union soldiers, this time in proper pine coffins. By then, "many of the bodies were only skeletons while on others the decomposed flesh still clung to the bones." According to one eyewitness, "wives, mothers, and sweethearts" who came to look upon their loved ones a last time "fainted or became hysterical" because "most [of the corpses] were unrecognizable."[16]

Meanwhile the killing went on. In January 1864, Henry C. Halleck complained to Ulysses Grant about "the public press" and congressional leaders "acting in the mistaken supposition that the war is nearly ended, and that we shall hereafter have to contend only with fragments of broken and demoralized rebel armies." Military leaders knew better. Grant—named general-in-chief of all Union armies on March 12—hoped that multiple coordinated offensives against the Confederates would finally allow the numerically superior Union forces to crush the rebellion. That could only be accomplished via wholesale death and destruction. President Lincoln heartily approved of the aggressive new strategy. As he explained, when it came to killing Confederates, "those not skinning can hold a leg." Grant assigned George Meade's Army of the Potomac to make Lee's army its objective. Similarly, he charged William T. Sherman, commanding the Army of Tennessee, with destroying Joseph E. Johnston's army in Georgia. Other smaller-scale campaigns were to occur simultaneously across the South, under the politically appointed generals Nathaniel Banks, Franz Sigel, and Benjamin Butler. However, Grant's multipronged offensive got off to a slow start, as Banks, Sigel, and Butler failed in their campaigns in Louisiana, the Shenandoah Valley, and outside Richmond, respectively.[17]

With the Union strategy in trouble, Confederate forces launched their own limited offensives in the winter and early spring of 1864. One of these, led by Nathan Bedford Forrest, still stands as one of the war's worst atrocities and an example of unwarranted wholesale killing. After raids across western Kentucky and Tennessee, Forrest arrived at Fort Pillow, north of Memphis, in April. Some 600 men, over half of them former slaves now serving in African American units, were inside. When the Federal commander declined to surrender on April 12, the Confederates attacked. As their defensive line collapsed, Union troops threw down their weapons and raised their hands or handkerchiefs in gestures of surrender. Southern soldiers, angry at the North's use of black troops, ignored these signals, heartlessly stabbing the unarmed Union men. Forrest's men killed 350 of the 600 men in the garrison. "The slaughter was awful," one Confederate sergeant wrote. "Words cannot describe the scene. . . . The poor deluded negroes would run up to our men, fall upon their knees, and with uplifted hands scream for mercy but were ordered to their feet and shot down." He recalled, "Their fort turned out to be a great slaughter pen. . . . Human blood stood about in pools and brains could have been gathered up in any quantity."[18]

Three weeks after that gruesome event and 300 miles away in northern Georgia, General Sherman put his 110,000 men in motion. Throughout the next two months, they moved south, largely avoiding costly direct assaults, preferring to outflank Johnston's 65,000-man army whenever possible. By July, Sherman was on the outskirts of the city of Atlanta, facing the Confed-

At the Battle of the Wilderness, some wounded soldiers never escaped the wildfire started by sparks from black powder rifles. From *Harper's Weekly*, June 4, 1864.

erate army under a new commander, John Bell Hood. Hood launched three failed assaults on Sherman's army that month, before digging in to withstand a siege. Though he had not destroyed the enemy army or captured Atlanta, Sherman, unlike most of his fellow generals, managed to avoid heavy casualties. Between May and September, Sherman suffered roughly 31,000 casualties (less than one-third of his force), while the Confederates lost 34,000 (more than half of their army).[19]

Sherman's successful maneuvers contrasted sharply with the brutal Overland Campaign that unfolded in central Virginia that summer. On May 4, the 120,000-man Army of the Potomac crossed the Rapidan and Rappahannock Rivers and engaged Lee's 65,000-man Army of Northern Virginia at the Battle of the Wilderness the next day. The thick cover and poor visibility turned the Wilderness into a new kind of death trap. Sparks from the guns set the dry brush on fire, and many wounded soldiers roasted alive as their pitiful screams filled the woods. As darkness fell on May 6, the fight had ended in a bloody draw. Nearly 17,500 Union troops had fallen (or been captured), with more than 2,200 of them killed. The Confederates lost approximately 11,000 men, with an estimated 1,500 killed. Refusing to retreat, Grant disengaged his army and moved it south. With grim determination,

Grant telegraphed Lincoln: "I intend to fight it out on this line if it takes all summer."[20]

Through a combination of Confederate quickness and Union lethargy, Lee was able to get ahead of Grant and block his progress at Spotsylvania Court House, a few miles south of the Wilderness. Two weeks of continuous battle culminated in the horrendous fight of May 12 at the Mule Shoe. Before finally disengaging and moving around Lee's right flank on May 21, Grant lost more than 18,000 soldiers, of whom more than 2,700 were killed. Lee suffered nearly 13,000 casualties, including approximately 1,500 killed in action. Two weeks later, Union and Confederate forces dug in with trenches and impressive fortifications northeast of Richmond. In the ensuing Battle of Cold Harbor, Grant launched a large-scale attack on June 3. One northerner described it as "a simple brute rush in open day on strong works." It did not go well. Waves of Union soldiers "fell as grass before the reaper—only swifter." In one hour of fighting, 3,500 Union troops fell, compared to only 700 of the well-protected Rebels. Grant later confessed, "I have always regretted that the last assault at Cold Harbor was ever made." During the two-week Cold Harbor campaign, the Union lost another 1,800 men killed, and over 9,000 wounded, while the Confederates lost fewer than 100 killed and 3,000 wounded.[21]

Grant decided to move the bulk of his army south to attack the vital rail junction at Petersburg, twenty-five miles below the capital, and on June 15 he ordered assaults that continued sporadically for four days. However, the wary bluecoats had seen the death and destruction that came from attacking fortified lines, and they failed to take the position. Union artillerist Charles Wainwright derided "this most absurd way of attacking entrenchments by a general advance in line." He summed up the troops' mental exhaustion: "It has been tried so often now and with such fearful losses that even the stupidest private now knows that it cannot succeed, and the natural consequence follows: the men will not try it. The very sight of a bank of fresh earth now brings them to a dead halt." During those four days of fighting, nearly 1,700 Union soldiers died, with more than 8,500 wounded. Confederates lost roughly 4,000 men, of which only 200 were killed. By the next day, Lee had moved his veterans into the undermanned defenses, compelling Grant to begin a siege of the city.[22]

Grant's frustration with his inability to destroy Lee's army led to the notorious engagement known as the Battle of the Crater. Here, new methods of killing and racial animus appeared side by side. The Union commander approved a plan to dig a tunnel under Confederate lines, fill it with explosives, and literally blow a hole in the defenses. After weeks of work, Union engineers detonated the first charges at 4:30 a.m. on July 30, 1864. The explosion immediately obliterated 278 South Carolinians who had the misfortune to be stationed just above the blast, but Confederates quickly rallied to plug the breach. Soon, many of Grant's attacking soldiers found themselves trapped in the newly created crater. When African American troops joined the assault, one Confederate officer recalled that the sight of the black men had the same effect on the Rebels "that a red flag had upon a mad bull." During intense, close-quarter fighting—mostly "done with bayonets and butts of muskets"—the southerners slowly gained the advantage. Another 500 Union soldiers perished and nearly 2,000 suffered wounds in the disaster. Confederates lost 1,500 casualties. As Federal troops realized the hopelessness of their situation and tried to surrender, Confederates on the rim of the Crater simply kept firing. As had happened at Fort Pillow three months earlier, Confederates had no interest in showing mercy to black soldiers. One Union officer saw "the rebs run up and shoot negro prisoners in front of me." A Mississippi soldier concurred: "Most of the Negroes were killed after the battle."[23]

When it came to death, the environment did not favor one race over another. African American soldiers began officially serving in the Union armies in the spring of 1863, and started seeing combat in large numbers in 1864. Just as a host of diseases ran rampant through black enlistment camps in 1863 as they had those of white troops in 1861, bullets and bayonets also killed black troops the same as they did their white comrades. After the Battle of the Crater, it did not take long for nature to begin its reclamation of the dead, regardless of skin color. One Maine officer recalled, "Men were swollen out of all human shape, and whites could not be told from blacks, except by their hair. So much were they swollen that their clothes were burst, and their waist-bands would not reach half-way around their bodies; and the stench was awful." A soldier observed that whether white or black, "the wounded were fly-blown, and the dead were all maggot-eaten." Burial details "had to lift them with shovels."[24]

The 1864 Overland Campaign illustrated the myriad ways soldiers could die in battle: by bullet, musket butt, bayonet, cannon, forest fire, drowning, or explosion. The Union defeat at the Crater ended three months of slaughter in the woods, fields, and fortifications of northern Virginia. From May 1 to August 1, Grant's army suffered approximately 70,000 casualties (including nearly 10,000 dead)—more combat casualties than the same sized army had suffered during the entire year of 1862, encompassing the bloody Penin-

Confederate dead at Spotsylvania, May 1864. Courtesy Library of Congress.

sula, Seven Days, Second Bull Run, Antietam, and Fredericksburg campaigns. Among the Confederates, perhaps 35,000 men fell. The siege lasted another eight months, costing tens of thousands of more casualties on both sides. Lee's administrative officers could no longer even count the killed and wounded, because constant campaigning prevented the routine roll calls. Casualty numbers became educated guesses.[25]

As the armies found new ways to kill, savvy businessmen developed a thriving trade in the preservation of human remains. Relatives sometimes journeyed to battlefields in search of dead loved ones, but as the geographic scope of the conflict expanded and casualties soared, families increasingly trusted that unpleasant task to others. As a result, nature's processes of decomposition had to be kept at bay long enough to allow the corpse to be shipped home (by rail or wagon) and arrive in suitable condition for viewing and burial. A well-preserved body provided the living with a final opportunity to visit with the deceased, an important facet of nineteenth-century grief and interment ceremonies. The war turned corpses into environmental commodities, harvested in the aftermath of battle, preserved for shipment, and marketed to distant consumers. Many of those consumers were Union families whose sons fell on southern soil.[26]

Those who dealt in the dead devised all manner of intriguing techniques to move bodies from the battlefield to burial grounds. Thomas Holmes, a "coroner's physician" practicing in New York City, emerged as one such innovator. On July 21, 1863, just weeks after Gettysburg, Holmes patented what he called "a coffin," but it more closely resembled a modern body bag. Made of natural rubber and shaped vaguely like a vase, with a wide opening at the top, the contraption came with a ventilated stopper that allowed for deodorants or perfumes to be placed in the bag without opening it. According to Holmes in his patent application, the "coffin" served to "facilitate the carrying of badly-wounded dead bodies that could not otherwise be removed . . . from the field of battle."[27]

Any dead man destined for distant burial went into some sort of casket for shipment via rail, wagon, or stagecoach. Companies already involved in shipping other parcels quickly adapted to the body business. The Adams Express Company and its subsidiary, Southern Express, initially offered wooden casket services to both Union and Confederate families. But those containers sometimes did little to slow decomposition. Families who opted for such services sometimes had to rely on local coach or railway agents who received the body and temporarily buried it on their property until the family could disinter the loved one and provide home burial.[28]

As any soldier assigned to burial detail could attest, the biggest problem with transporting the dead—whether to a makeshift grave or elaborate interment—was the sickening odor that set in during the early stages of decomposition. Metal coffins, thought to be better for preservation, were in use before the war. Caskets with separate bottom chambers that could be lined with ice to cool the body or charcoal to absorb odors also became popular as the fighting intensified. In Hartford, Connecticut, in 1863, an advertisement for "new patent body preservers" assured first-time purchasers that "no ice comes in contact with the body," leaving the deceased "dry and perfect, at half the cost of any other method."[29]

That same year, *Scientific American* gave notice of a new "air-tight deodorizing burial-case," patented by G. W. Scollay of St. Louis. Scollay employed a sealant to make sure his wooden casket could "retain the fluids

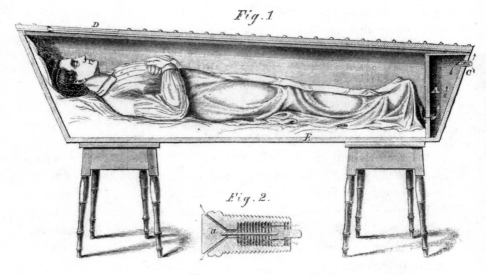

"Scollay's Patent Air-Tight Deodorizing Burial-Case,"
Scientific American 8, no. 9 (February 28, 1863).

arising from the decomposition of the body." The coffin also had a separate "deodorizing compartment or chamber" that employed chemicals to dissipate the scent of death. An "elastic valve," triggered by various gases escaping from the corpse, opened the small chamber periodically and allowed the treated fumes to escape "into the outer air, destitute of the slightest disagreeable odor, and perfectly free from contagious or infectious properties." A glass plate at one end of the burial case allowed mourners to see the face of the deceased. Allegedly, even the "lapse of months [did] not render the body offensive to anyone." It was a timely innovation, *Scientific American* noted, because "desolation and grief exist in almost every home in the land." Such technology usually proved expensive. Though the notice about Scollay's invention did not list a price, some of the more elaborate caskets could run upward of fifty dollars plus the price of shipping the body. At a time when soldiers made only thirteen dollars a month, Scollay's casket was a luxury most nineteenth-century families could not afford.[30]

For wealthier families, chemical preservation of the body itself, commonly called embalming, provided the best chance of getting human remains home intact. Thomas Holmes, inventor of the body bag, had long studied European embalming techniques and by the time the war began, Holmes perfected his own embalming fluid. The concoction promised to do the job without dangerous chemicals or poisons, though it apparently contained arsenic. Holmes received nearly nationwide attention in 1861 when he embalmed the body of President Lincoln's friend Elmer Ellsworth, killed in a confrontation with a southerner at Alexandria, Virginia. Viewing Ellsworth's body, mourners, including the first lady, found it almost lifelike. As the body count rose during the following years, Holmes trained other undertakers in the macabre art and sold his patented fluid for three dollars a bottle. Working out of Washington, Holmes preserved some 4,000 fallen soldiers, charging relatives $100 per body. As Faust notes, "The war made him a wealthy man." After training with Holmes, some embalmers took their skills directly to the battlefield, offering on-site service to those who could pay, though prices and results varied widely from one practitioner to another. Embalming proved less common in the South, but several firms in Virginia offered funeral arrangements that included chemical preservation.[31]

Perhaps nothing illustrates the wartime commodification of human remains more than the advertising associated with embalming and its ability to arrest natural processes. Nearly all undertakers using the process promised to halt or at least postpone nature's inevitable reclamation of the body. Holmes showed off his talents by recovering unidentified corpses from battlefields, embalming and dressing them in appropriate clothing, and putting the bodies on display along the streets of Alexandria and Georgetown. Presumably, passersby could judge for themselves the extent to which Holmes had eliminated the bloating, discoloration, and odor of decomposition. Print ads for other embalmers in the nation's capital proudly proclaimed that "Bodies Embalmed by US NEVER TURN BLACK! But retain their natural color and appearance." Whether such claims increased business remains an open question. Given the high cost of preservation and shipment, numerous families, especially on the Confederate side, had to leave their relatives where they fell and simply allow nature to take its course.[32]

For many soldiers the fear of a debilitating wound proved as profound as the horror of becoming a cadaver. When the war began, neither the Union nor the Confederacy had developed a large medical corps, because few thought the war would last long. When South Carolina seceded in December 1860, only 114 doctors served in the Union army, and 24 of them eventually resigned and joined the Confederacy. Most military doctors had served at small outposts before the war. Few had ever seen a gunshot wound, much

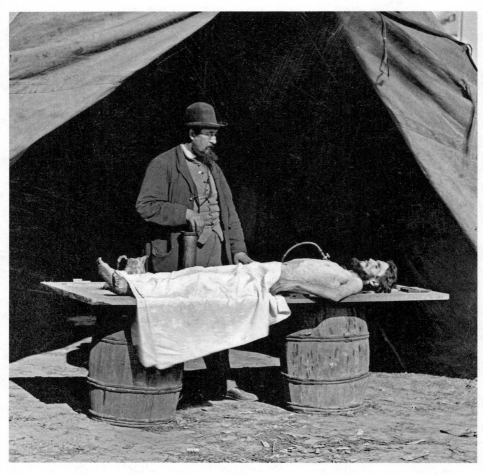

Richard Burr, an embalming surgeon in the Union's Army of the James, demonstrating the procedure on a dead soldier. Courtesy Library of Congress.

less the kind of carnage that occurred in major battles. In 1861, state authorities were so eager to enlist men into service that they signed up volunteer surgeons who did not have medical degrees.[33]

Though Civil War soldiers wielded a variety of weapons, most wounds resulted from gunshots. Smoothbore muskets firing round projectiles had been common in 1861, but by the fourth year of the war, nearly every soldier carried a rifle loaded with the minié ball. Named after French army officer Claude-Étienne Minié, who helped develop it in the 1840s, the minié ball was a conical bullet made of soft lead. It was large (.58 caliber) and heavy (two ounces). Civil War rifles fired bullets that traveled 963 feet per second—a low muzzle velocity. (By way of comparison, M-16s used during the Vietnam War fired a bullet one-third as large that traveled 3,250 feet per second.) While modern bullets have a metal jacket, leaving a clean wound, the large, heavy, soft lead minié ball flattened and tumbled when it hit a human body, tearing away hunks of flesh and shattering bones into spicules. With nearly 40,000 wounded men on the battlefields of the Wilderness and Spotsylvania alone, surgeons in both armies worked nonstop to save as many lives as possible. Unlike early in the war, when the military medical establishment was woefully unprepared for battle casualties, by 1864, those systems had grown much more efficient. As a result, a comparatively large number of men lived, albeit often permanently disabled.[34]

The armies also lacked viable procedures for evacuating and treating wounded men quickly. In 1861, with no dedicated medical corpsmen, members of military bands that accompanied the soldiers—more easily spared because they were not directly involved in the fighting—often served as stretcher-bearers. Carrying a wounded infantryman on an improvised blanket-stretcher was no simple task; it took four men holding corners to transport the injured soldier to safety. As Georgia soldier-turned-poet Sidney Lanier recalled, "Easy walking is desirable when each step of your four carriers spurts out the blood afresh, or grates the rough edges of a shot bone in your leg." At many battles early in the war, the Union armies had to leave hundreds of wounded soldiers untended because the medical corps lacked the means to move them.[35]

Like the dead, wounded survivors entered new ecological relationships with other organisms that inhabited the battlefield. One of the strangest of those interactions might have occurred following the Battle of Shiloh in April 1862. After that bloody encounter, wounded soldiers from both armies remained on the cold, rain-soaked battlefield for two days awaiting transport to hospitals. During the night, some of the men noticed that their wounds seemed to glow with blue phosphorescence. Oddly enough, once off the battlefield, those who reported illuminated wounds appeared to heal faster than others in the same hospitals. As a result, the phenomenon came to be known as "Angel's Glow."[36]

At the time, soldiers and their families regarded this "Angel's Glow" as a form of divine intervention that kept injured men safe, but it might also have a scientific explanation. Exposed wounds quickly attracted insect larvae and those larvae, in turn, attracted predatory nematodes, tiny worms that live in soil. Modern researchers know that, as nematodes subdue their prey, they release bacteria into the larvae. Some of those bacteria are bioluminescent; they glow. Once inside the larvae, both bacteria and nematodes feed on the host until it dies. In the process the bacteria eliminate other competing microorganisms, including some of those that can cause infection in humans. Theoretically, both the nematodes and the glowing bacteria made their way into the open wounds of men who fell at Shiloh. In other circumstances, the human-friendly microbes likely would not have survived long enough to aid the victims. The glowing bacteria cannot live at normal body temperature. Thanks to the rain, cold, and prolonged exposure, however, it seems possible that the injured troops also suffered from mild hypothermia, a condition that lowered body temperature just enough to allow the helpful bacteria to enter the wounds. The wounds glowed as the phosphorescent bacteria and their nematode allies took up residence and eliminated other microorganisms that might cause infection. Once off the battlefield and free of hypothermia, the soldiers no longer glowed, but their relatively clean wounds healed faster and with fewer complications than injuries not exposed to the bizarre phosphorescence.[37]

Angel's Glow appears to have been an isolated phenomenon, confined to the peculiar conditions of Shiloh in early spring. Other forms of insect infestation proved far more common. Any open gunshot wound quickly attracted blowflies and houseflies that deposited eggs in the exposed flesh. In the warm humid conditions common in summer campaigns, the eggs usually hatched within twenty-four hours. Injured soldiers left out overnight in warm weather often awoke to find their wounds teeming with fly larvae. At the First Battle of Bull Run, a rifle bullet smashed Union private Alfred D. Whitehouse's right arm, and he lay on the battlefield without treatment. By the time Union medics finally got to him six days after the battle, his "shattered arm was full of maggots." Hospitals also became breeding grounds for maggots. Nurses considered the larvae a disgusting nuisance, while patients complained bitterly of "the annoyance created by the continual sensation of crawling and irritation which it occasions." Northern doctors used various methods to keep the larvae at bay, including killing them with chloroform and carefully cleaning wounds every two to three hours.[38]

Had Union physicians known a bit of history, they might not have been so quick to purge the repugnant larvae. Maggots have been linked to wounds and battlefield medicine at least since biblical times when Job's "broken and loathsome skin" became "clothed with worms." Some evidence suggests that Mayan doctors saturated new bandages in cow's blood and left the dressings outside to attract maggots. Only then were the wormy cloths applied to battle wounds. In 1557, French royal physician Ambroise Paré noted that, after the Battle of Saint Quentin in northern France, soldiers with maggots in their injuries recovered faster than those who escaped infestation. Napoleon's doctors reported similar findings during his Egyptian campaign in 1799. However revolting they might sound, such observations had a sound basis in science. Though modern physicians do not fully understand the process, researchers have discovered that maggots consume only the dead tissue around a wound, leaving healthy flesh alone. As they feed, the larvae seem to secrete some sort of "antibacterial agent" that deters a variety of pathogens. Enzymes in a maggot's digestive apparatus also break down bacteria. In addition, maggot activity raises pH in and around the wound, which encourages healing.[39]

Confederate doctors at first shared the common revulsion toward fly larvae, but many southern surgeons lacked the necessary personnel and supplies to keep their patients' wounds clean and pest-free. Such conditions resulted in an accidental semicontrolled experiment with maggots at a Chattanooga prison stockade in December 1863. Confederate surgeons tending the wounded Rebel prisoners there found themselves without suitable bandages. In short order, infections set in and nearly all of the injuries became maggot-infested. Meanwhile, better-equipped Union physicians in Chattanooga hospitals scrupulously cleaned and bandaged northern soldiers' wounds to deter the larvae. To the surprise and delight of the Confederate doctors, their patients healed quicker and with fewer complications than their northern counterparts. Joseph Jones, a high-ranking Confederate surgeon, witnessed similar results several more times during his career, explaining that "a wound which has been thoroughly cleansed by maggots heals more rapidly than if it had been left to itself." At least one southern physician intentionally employed fly larvae in wound treatment. Like others before him, J. F. Zacharias, a surgeon at a Danville, Virginia, prison, observed that maggots "would clean a wound much better than any agents." Acting on such observations, he apparently became the first American doctor to prescribe maggots for the patients in his care. "I am sure," he wrote, that "I have saved many lives by their use, [as the injured] escaped *septicaemia*, and had rapid recoveries."[40]

On the Union side, any anecdotal evidence that maggots might be beneficial appears to have been lost or ignored. Given their prevailing belief that cleanliness aided recovery and their better access to bandages and other medical supplies, northern physicians could hardly be expected to embrace the technique. It took decades for the intentional use of maggots to gain credence in the medical community. Modern doctors, however, have found the larvae effective in treating certain injuries that are slow to heal due to diabetes or other chronic ailments. Today, "maggot debridement therapy" is an accepted (if not so widely used) technique for wound management.[41]

Rats also infested Civil War camps and hospitals, provoking similar disgust among most medical personnel. Yet rats, too, could aid physicians. Phoebe Pember, a nurse and administrator at a Confederate hospital outside Richmond, witnessed one such episode. Examining a Virginia soldier with a foot injury, doctors discovered that the infected wound had developed a mass of dead tissue that impeded healing. According to Pember, "Surgeons feared to remove the putrefying matter, thinking it was connected to the nerves of the foot and lockjaw might ensue." Where doctors feared to tread, the hospital's resident rodents rushed in, consuming the dead tissue and leaving the wound "washed clean and looking healthy," prompting the soldier to display "the foot with great glee." According to Pember, "The rat surgeons could have passed the board [of examiners that certified doctors]."[42]

While Confederates benefited from such unusual methods, the North successfully revamped its medical corps to improve traditional methods of treatment. The reorganization began in 1862 after the Second Battle of Bull Run. With 3,000 Union wounded left on the field, Surgeon General William A. Hammond hired private contractors with over 200 carriages and wagons to retrieve the injured. Many of the new ambulance drivers left the wounded where they lay and stole the medicinal liquor supplies instead. In disgust, Hammond wrote the secretary of war, noting that a week after the battle some 600 wounded still lay on the battlefield. "Many have died of starvation," he wrote, and "many more will die in consequence of exhaustion, and all have endured torments which might have been avoided."[43]

That shameful performance shocked the Union medical establishment into action. Jonathan Letterman took over as medical director of the Army of the Potomac in July 1862. Appalled at the treatment of the sick and wounded, he immediately began a series of comprehensive reforms. He developed a professional and dedicated ambulance corps, improved evacuation techniques, and established field hospitals at the front as well as a series of general hospitals in rear areas to treat the more seriously wounded. Though the War Department did not immediately mandate Letterman's reforms, his protocols spread as the war continued. Additionally, armies on both sides also hired a large number of surgeons to treat the ever-growing number of sick and wounded soldiers. By 1865, the Federal service had approximately 11,000 surgeons, or one doctor for every 133 soldiers. The South could only reach a ratio of one doctor for every 324 men. Letterman's system meant that wounded Union soldiers received quicker and better treatment, enhancing their chances of survival. During the first year of the war, the death rate from wounds had been over 25 percent. It had decreased to 15 percent in the second and dropped to only 9.5 percent in the third.[44]

As the Overland Campaign of 1864 began, the army's new medical director, Thomas A. McParlin, attempted to utilize these new policies to save lives. Early on, McParlin established a major hospital at Brandy Station, a few miles in the rear of Union lines. He prepared nearly 500 ambulances to transport wounded from the front to the hospital complex. It was not enough. During the Battle of the Wilderness, surgeons had to procure an additional 320 wagons to accommodate the vast numbers of wounded that had to be hauled from the field. As the ambulances headed toward Brandy Station, Grant changed his orders and moved south around Lee's army to Spotsylvania. The road quickly filled with supply wagons heading to the army, forcing the ambulances to divert to Fredericksburg, a more distant destination serviced by worse roads. The medical authorities had not prepared Fredericksburg to receive wounded soldiers, so personnel there had few supplies. In short order enormous numbers of wounded from the Spotsylvania battles overwhelmed the medical facilities.[45]

Getting to an ambulance alive did not guarantee survival. Though Angel's Glow, blowfly maggots, and rodents might aid in healing, gunshot wounds also brought soldiers into contact with other, far more harmful organisms that flourished on battlefields and in hospitals. Every surgeon feared gangrene, a general term for bacterial infections that sometimes set in with frightening rapidity. Telltale signs of gangrene included inflammation and oozing pus, usually classified as either "laudable pus" or "malignant pus." Given the nomenclature of the time, both descriptors might indicate *Staphylococci* or *Streptococci* bacteria, either of which could cause a serious infection. A third bacterium, *Clostridium,* proved even worse. It resulted in a virulent infection known as gas gangrene that aggressively destroyed both skin and muscle. Gas gangrene could often be identified by the foul stench emanating from the infected area.[46]

Because the destructive minié ball frequently shattered bones in arms and legs, the best available defense for gangrene was amputation. Union doctors performed nearly 30,000 amputations during the war; estimates suggest a similar number for Confederate surgeons. Amputation did not always prove effective in keeping infectious bacteria at bay. By modern standards, the mortality rate for Union amputees remained high at 26.3 percent. Even so, the death rate was less than half that of civilian amputees in America and Great Britain when the war began. As the war progressed and surgical

Field hospital at Savage Station, Virginia, June 27, 1862.
Courtesy Library of Congress.

skills improved, doctors discovered that the quicker a damaged limb could be amputated, the greater the likelihood that a wounded man might survive. In one study of nearly 24,000 Civil War amputees, those who had surgery within the first forty-eight hours of their wound had a 23.9 percent fatality rate, while those who had surgery more than forty-eight hours after a wound suffered a 34.8 percent fatality rate.[47]

Due to the prevalence of amputation and infection, soldiers, nurses, and civilians often disparaged military hospitals. Louisa May Alcott, who would pen *Little Women* after the war, wrote of one such facility in the nation's capital: "A more perfect pestilence box than this I never saw—cold, damp, dirty, full of vile odors from wounded, kitchens, and stables." As a nurse, she developed typhoid fever and ingested so much calomel that she lost most of her hair and teeth and suffered from shooting pains in her arms and back for the rest of her life. Hospitals quickly became known as places for soldiers to avoid. One Georgia man voiced a dread that many shared when he stated, "I never want to go tho it may fall my lot to go thare and dy."[48]

Despite the common perception among soldiers and the public, Civil

War hospitals continued to develop more efficient care as the war progressed. In 1861, the nation's prewar army had only one forty-bed hospital; by war's end, Union and Confederate authorities had established 400 hospitals with a total of over 400,000 beds for patients. Washington had one hospital in 1861, but fifty by 1865. In Richmond, the Confederate medical bureau established Chimborazo hospital—the largest hospital complex in the world—with 150 buildings and 8,000 beds spread over 125 acres. By the end of the war, Chimborazo had treated over 76,000 patients. More than 1 million Union soldiers were treated in a hospital during the war, and fewer than 10 percent of those patients died during the war. After the war, European armies copied the American hospital system and modeled their medical care on the American practices developed during the conflict.[49]

In addition to the thousands killed and wounded, roughly 18,000 soldiers on both sides surrendered during the Overland Campaign battles, and most spent the rest of the conflict in captivity. Over the course of the war, nearly 144,000 Union and 215,000 Confederate soldiers spent time in prison camps. These institutions were frequently grim repositories of human misery, where death and disability ran rampant. Approximately 30,000 Union soldiers and 26,000 Confederate troops died in prisons, roughly 7.5 percent of the war's dead. Many others emerged sickly, weak, or permanently disabled.[50]

Many Confederate prisoners of Virginia's 1864 campaigns found themselves shipped to the newest Federal prison in Elmira, New York. The forty-acre complex was developed in May 1864 on the site of a former military training center. The first Confederate prisoners arrived on July 6, and within two months over 10,000 filled the camp designed to hold only half that number. In mid-August, the director of northern prisons ordered rations reduced to bread and water in retaliation for the severe conditions in southern prisons. Malnutrition and polluted drinking water ushered in giant outbreaks of scurvy and diarrhea, while pneumonia soon followed. Prisoners referred to their facility as "Hellmira," and many died of starvation, disease, and exposure. In order to handle the large number of deaths, authorities contracted with local workers to establish efficient mortuary and burial services, prompting one southerner to remark, "The care of the dead was better than that bestowed on the living." During its year of existence, slightly more than 12,000 Confederate prisoners entered Elmira's gates, and nearly 3,000 found permanent rest in local cemeteries. Though it was only one of twenty-seven Union prisoner of war camps, Elmira had the highest death rate of any northern military prison.[51]

No prison was more notorious than the Confederacy's Camp Sumter, near Andersonville, Georgia. Opened in February 1864, as a 16.5-acre rectangular expanse, enclosed by a wooden stockade, the prison was originally intended to hold 10,000 soldiers. By June 1864, Confederates had enlarged it by ten acres but still could not adequately accommodate the massive influx of prisoners taken in the spring battles. By August, over 33,000 inmates resided there. When Capt. Robert Kellogg of the Sixteenth Connecticut regiment arrived in May 1864, he recoiled at what he encountered: "As we entered the place, a spectacle met our eyes that almost froze our blood with horror, and made our hearts fail within us. Before us were forms that had once been active and erect; — *stalwart men*, now nothing but mere walking skeletons, covered with filth and vermin." The shock of the sight prompted some of the Connecticut soldiers to exclaim, "Can this be hell?"[52]

With no constructed lodgings, the prisoners had to create their own shelter out of coats and tent cloth. The sole source of water within the stockade — a branch of the legendarily misnamed Sweetwater Creek — quickly became filled with human waste yet served as the primary source for bathing and drinking water for most soldiers. Bacterial dysentery spread at lightning speed, and as a result one Massachusetts prisoner wrote in June, "The stench which pervades every part of the camp is well nigh unendurable." The inmates received such scanty portions of food that some men even scoured for scraps or remnants of undigested food among the human waste in the latrines. Lack of vegetables meant scurvy dominated the camp, while the "torrid heat" of the Georgia summer baked the sick and malnourished soldiers. A Confederate medical officer reported on the terrible conditions in the post hospital: "Millions of flies crawled over everything and covered the faces of sleeping patients and crawled down their open mouths and deposited their maggots in the gangrenous wounds." One soldier wrote of "the sick & dying all around [that] make this a scene of horror which will ever be vivid in my memory." Over 45,000 men entered Andersonville; nearly 13,000 died, mainly from malnutrition and bacterial related diseases. In the first half of September, an average of 102 men died every day. One Confederate physician described it simply as "Hades on Earth."[53]

The appearance of the survivors sometimes shocked civilians. When Walt Whitman saw some of the Andersonville prisoners in February 1865, he was horrified. "Can these be *men* — these little, livid, brown, ash-streaked, monkey-looking dwarfs?" he asked. "They lay there, most of them quite still, but with a horrible look in their eyes and skinny lips — often with not enough flesh to cover their teeth." Whitman, who had witnessed a lot of suffering during his stint as a wartime nurse, proclaimed, "Probably no more appalling

Andersonville prison, from a photo taken August 17, 1864, while the facility housed 33,000 men. Over 100 prisoners would die every day in early September 1864. Courtesy Library of Congress.

sight was ever seen on this earth." One Chicago medical examiner described starved Andersonville survivors thusly: "Their arms and legs look like coarse reeds with bulbous joints. . . . Their faces look as though a skillful taxidermist had drawn tanned skin over the bare skull, and then placed false eyes in the orbital cavities." One Wisconsin soldier recorded, "I was so changed in appearance that nobody knew me, not even my mother." He confessed, "I almost wished then that I had died at Andersonville."[54]

Psychological scars caused survivors of the war, not just former prisoners, to experience nightmares for years. Some men repeatedly tried to scratch an itch in phantom limbs. Others fell into depression. While the "empty sleeve" became a badge of honor after the war—a testament to their service and sacrifice—wounded soldiers also knew that such disabilities carried a stigma in nineteenth-century society. Disfigured soldiers feared the shame they faced at home and worried that their disability might affect their work or relationships. One Michigan soldier lost his fiancée when she learned he had lost his arm. "I offered her my other hand / Uninjured by the fight; Twas all that I had left," he penned. "'Without two hands,' she made reply, 'You cannot handsome be.'"[55]

Returning wounded soldiers—alive but unable to perform physical labor—faced other difficulties providing for themselves and their dependents. Disabled veterans transitioned from occupations that required manual labor toward clerical work. In one study, only 3 percent of soldiers had been clerks at the time of their enlistment, but 23 percent held that occupation in 1870. Because the expanding Federal government needed a veritable army of civil servant clerical workers after the war, the disabled veterans filled those positions in large numbers. New laws granted wounded soldiers preferential treatment in the hiring process. Usually such employment did not provide much satisfaction to those soldiers who exalted the economic independence of farming or artisanal work. "Disabled veterans who worked as clerks," historian Jalynn Padilla argues, "did not celebrate their work as a step up the occupational ladder."[56]

Civilians struggled with how to process and accept so many disabled soldiers returning to peacetime society. They could honor those who gave the last full measure of devotion easily enough but sometimes failed to appreciate those who had sacrificed and suffered but survived. As one Union veteran noted, "In heaven's name, while we remember the dead let us not forget the living." This was not a uniquely northern phenomenon. Roughly 200,000 disabled or gravely ill Confederates returned home after the war, often to an ambivalent reception. In a self-published memoir, one Texas soldier wrestled with his wartime service and the lack of respect wounded veterans received in the postwar period; he strikingly titled his autobiography, *Harder Than Death*.[57]

While the inability to maintain independence as farmers and society's disregard for their handicaps distressed wounded veterans, having to seek charity or government pensions emasculated them. By 1890, nearly 750,000 soldiers or their dependents received a pension from the Federal government. Beginning that year and continuing for the next two decades, a quarter of Federal expenditures supported pensioners. Similarly, in some southern states, pension obligations consumed more than 50 percent of state budgets in the years after the war. So much of the Federal budget was going to wounded veterans that a public backlash ensued. One Chicago newspaper editor claimed that the "army of pension beggars" proved "more dangerous enemies of the nation than undisguised rebels." A northern minister declared that the pensioners led "a raid upon the treasury." Government officials believed that perhaps as many as 33 percent of pension claims were either outright fraudulent or greatly exaggerated. As a result, the requirements to prove one's disability became onerous and humiliating for genuinely injured soldiers. One Indiana veteran mordantly quipped that an army physician took "less than fifteen minutes" to clear him for military service in

1861, but it took pension examiners "more than fifteen years" to determine that he could not physically support himself.[58]

Moreover, dead and maimed soldiers could not return to farming their fields. This affected land use in the North but more so in the South. In the years immediately following the Civil War, total acres of cultivated land decreased throughout the South. In the eleven Confederate states, between 1860 and 1870, the amount of improved land decreased by nearly 10 million acres. The lack of manpower also led to reductions in the harvesting of food crops in the years following the war. Much of the South faced destruction from battles or armies on both sides, and damaged land and farm equipment certainly played a role in this reduction of tilled land and foodstuffs. With slaves now freed by the war, that labor source also had to be replaced. But the decimation of the white working force through death or disability from war accounted for a significant part of the reduction. One out of every four southern soldiers died in the war, and likely a similar percentage returned home physically unable to perform much or any of the labor they had before it.[59]

The same battles that took such an enormous toll on human bodies also had important implications for the physical landscape. Downed fences, ruined agricultural fields, dead livestock, and the absence of some once conspicuous organisms — namely, white men — told only part of the story. Wherever armies fought and soldiers marched, the terrain itself often bore witness to the unparalleled carnage. And like maimed men, the terrain would require time to recover.

Six

TERRAIN

FALL 1864–SPRING 1865

It was something residents of the Virginia highlands probably thought they would never see: 600 black Union soldiers, most of them on horseback, slowly making their way across the southwestern corner of the state. They were members of the Fifth U.S. Colored Cavalry (USCC), a newly formed regiment that had recently joined 5,000 white soldiers commanded by Gen. Stephen G. Burbridge in Kentucky. The general had served as the Bluegrass State's military governor, where his ruthless treatment of Confederate sympathizers had earned him the nickname "Butcher Burbridge." He hoped the Virginia mission might help put those days behind him and improve his standing with the Union high command. The path Burbridge took into the Old Dominion proved treacherous; eight men and horses perished during one mountain crossing when they fell off the path. Worse, the rough terrain and Confederate guerrillas kept Burbridge from contacting his superior officers. As fate would have it, he missed an order from Gen. William T. Sherman that would have cancelled the risky mission.[1]

Burbridge had never been enthusiastic about using black troops, and he made sure that the best rifles and ammunition went to his white soldiers. The white men under his command openly ridiculed the African Americans, wondering aloud if the Fifth USCC would flee when the fighting started. Finally, on October 2, 1864, Burbridge and his army reached their destination, a low, flat valley in Virginia's Smyth County. Surrounded by higher ridges and flanked to the west by the North Fork of the Holston River, the valley had been the target of three previous Federal expeditions, all of which had failed.[2]

On that October day, Burbridge found the site defended by some 2,800 Confederate troops. Most of the Rebels had dug in on the surrounding ridges, where they manned a ring of forts, gun emplacements, and infantry trenches. From their defensive positions, the Rebels fired mercilessly on Union soldiers as they tried to cross the North Fork at the main ford. The surrounding gorges and ridges echoed with the sounds of battle. The Fifth USCC and the Twelfth Ohio Cavalry assaulted the Rebels' right flank, advancing up an imposing ridge held by two entrenched Confederate brigades. Meanwhile, more of Burbridge's units attacked the center of the Rebel fortifications.[3]

By all accounts the black troops fought courageously. The Fifth USCC and their comrades dismounted and advanced 800 feet up the ridge in the face of withering fire and drove the Confederates back to their fort on the ridge crest. The other Union forces did not fare as well, though, and no unit came close to overwhelming the enemy. By day's end, the battle had exhausted Burbridge's men and left them critically short of ammunition. Under cover of darkness, the Federals retreated, having suffered 300 casualties, compared to the Confederates' 160.[4]

Had the killing stopped there, the assault might be remembered only as a minor encounter. Instead, over the next few days, victorious Confederates from Tennessee searched the battlefield for wounded black soldiers and systematically executed any that still drew breath. In one instance, according to a Kentucky soldier who witnessed it, "a boy, no more than sixteen years old, with a pistol in each hand," entered a log cabin and killed at least seven wounded members of the Fifth USCC who had sought shelter there. A few days later, the Confederate guerrilla fighter Champ Ferguson walked into an infirmary at nearby Emory and Henry College and fatally shot two or more black patients and a white officer. The exact number of African Americans executed in the aftermath of the battle remains uncertain. Best estimates suggest forty-six racially motivated murders, ranking the massacre among the worst Civil War atrocities.[5]

Like most war crimes, the murders of black soldiers in southwestern Virginia reflected the worst of human nature: ambition, poor leadership, frustration, racism, and rage. What might be missed, though, is that neither Burbridge's troops nor the Confederate defenders would have been in that isolated valley that October had it not been for the region's unique terrain. Thanks to the area's peculiar geologic history, the valley sat atop a huge cache of a mineral that in late 1864 might have been more valuable than gold, at least to the Confederacy. For nearly a century, residents had extracted it by the ton and sold it across the South. The earliest white settlers even named the valley's only substantial town for it. They called the place Saltville.

In the nineteenth century, salt (sodium chloride) provided far more than flavoring for food. Salt served as the primary preservative for meat, fish, and some vegetables; it played a crucial role in the preparation of both butter and cheese. Domestic animals required it as a dietary supplement. Salt was essential in leather production and in the preparation of various medicines and cleaning compounds. No army could survive for long in the field without it. Fortunately, sodium chloride is one of nature's most common substances. One readily accessible source is seawater, from which salt can be extracted through evaporation. Such operations, driven by energy from the sun or from wood-fired salt kettles, sprang up from the earliest days of colonization in New England. The mineral also occurs in briny inland waters such as the Kanawha springs near Charleston, West Virginia. Salt from Kanawha helped preserve Cincinnati-processed pork that made its way into the South in the antebellum years. During peacetime, production there and at other smaller operations in Alabama, Louisiana, and Kentucky waxed and waned, depending on the market price for salt.[6]

Saltville drew its name from a related but far more viable source of sodium chloride deposited near the surface eons ago. At some point, maybe 350 million years past, the continent now called North America had a climate much like that along the modern Persian Gulf. As in any dry coastal climate, tides left seawater trapped in shallow depressions along the shore. As that water evaporated, salt and various other minerals began a chemical process known as precipitation that turned them into solid deposits. As such, the minerals then became trapped in the rock formations below. One such geologic feature, known as the Maccrady Formation, lay beneath Saltville.[7]

By the time Native people moved into the region (at least 15,000 years ago), they found briny springs and seeps where natural precipitation brought sodium chloride to the surface. When Europeans settled the valley and began to exploit the resource, the salty water table became both a blessing and a curse. Salt could be procured by evaporation, but tapping into the deposits via mines proved difficult because water frequently flooded the shafts and tunnels. By the late 1840s, the Saltville works had developed a method that involved drilling wells to draw up the salt-laden water into large metal containers. Laborers then boiled the water in smaller iron kettles housed in large open wood-fired furnaces. Each furnace heated 80 to 100 such kettles. Every thirty-two gallons of water produced one bushel (fifty pounds) of salt, "as white and pure," one observer noted, "as the driven snow." A branch line of the Virginia and Tennessee Railroad, which opened in 1856, annually carried some 10 million pounds of salt to half a dozen other states. For a variety of

In December 1864, Union forces destroyed the works at Saltville, Virginia, temporarily shutting down the Confederacy's last major source of salt. From *Harper's Weekly*, January 14, 1865.

reasons, production declined in the years before the war, but the onset of hostilities caused salt prices to soar, and the Saltville operations agreed to provide the Confederate Army with 22,000 bushels of salt per month.[8]

From 1861 to 1864, Union operations took out or severely restricted production at salt refineries in West Virginia, North Carolina, Georgia, Alabama, Kentucky, Arkansas, and on the Virginia coast. Together with the Federal blockade that interdicted shipments of salt from Great Britain—from which southern states had imported 3 million bushels in some years before the war—those attacks created a critical shortage of sodium chloride in the Confederacy. The South, which had used an average of 9 million bushels (450 million pounds) of salt a year before the war, could only obtain a fraction of that after 1861. By the fall of 1864, Saltville provided the bulk of the South's needs. The works expanded to include at least thirty-seven furnaces, housing 2,800 kettles and generating 4 million bushels of salt.[9]

That Saltville held out for so long owed much to its isolation and geologic history. The Maccrady Formation is part of the much larger Ridge and Valley geologic province. The Ridge and Valley had its origins in the Paleozoic Era, perhaps 500 million years ago as movement of Earth's tectonic plates caused continents to collide, driving up the spine of the Appalachians. For at least 200 million years, since North America became a distinct continent, the mountains have been exposed to erosion via wind and rain. As a result, different types of rock broke down at different rates. Limestone, one of the ancient ocean deposits driven toward the surface by sporadic periods of upheaval, wore down relatively quickly, creating long flat valleys. Sandstone, another rock from the sea floor, weathered more slowly, eventually helping to form hills and mountains visible today. Any Union troops moving on the region from Kentucky had to wind their way through the mountains along routes, mostly in the limestone valleys, well known to the Confederates. In contrast, southern defenders took to the high, erosion-resistant ridges, forcing the enemy to attack uphill. Even the valiant efforts of the Fifth USCC were not enough to overcome that long geologic history, which southerners turned to their advantage in October 1864.[10]

Centuries of human habitation had also shaped the contours of the Saltville battlefield. The salt furnaces consumed firewood at an alarming rate, even by nineteenth-century standards. Before the war, one agent estimated that every 100 bushels of salt required three cords of hardwood. Using that figure, the 4 million bushels produced in 1864 required some 120,000 cords, the equivalent of a stack of wood 4 feet high by 4 feet wide by 181 miles long. The insatiable demand inevitably led to increased cutting of the region's woodlands. Illustrations of the Saltville works from *Harper's Monthly Magazine*, published in 1857, show a valley nearly devoid of large trees, with the bulk of land given over to the saltworks, crops, and pasturage. The cost of wood also suggests depletion. In and around Saltville, prices for firewood soared. In January 1863, one buyer noted that a cord of wood could be swapped for 2.5 bushels of salt. Since salt sold for $25 a bushel, that cord of wood had an actual value of $62.50. By the standards of the day that was an exorbitant price. Two years later, in Union-occupied Savannah, a cord of firewood cost only $10; in Lexington, Kentucky, it went for $30. Even in Richmond, where overcrowding, shortages, and wartime inflation had driven prices for basic amenities to unheard-of levels by 1865, one could purchase firewood for $50 a cord.[11]

With prices inflated and viable wood in short supply, workers began harvesting trees from forests about ten miles south of Saltville, along an east-west line of the Virginia and Tennessee Railroad. Traffic jams ensued as train cars loaded with firewood interfered with salt shipments from the works. On cutover land devoted to pasturage, cattle helped keep regrowth to a minimum, eventually creating grassland. Along the railroad, selective cutting of the best trees probably left some timber standing. In addition, hardwood stumps often sprouted shoots, creating a low tangle of dense foliage in only a few years.[12]

The overall effect was to create a patchwork of forests, cutover brushy woodlands, and pastures for miles around the saltworks. As a result, Union troops frequently had to cross semiopen terrain in 1864. Well aware of why the Federals had targeted the valley in the first place, southern soldiers taunted the clearly visible enemy yelling, "Come right up and draw your salt" as the Union men crossed the stump-strewn fields. After the battle, a Confederate soldier came upon a dead Union officer whose head had been caved in by a projectile. According to one witness, the soldier "took a handful of salt from his haversack and threw it into the cavity in the officer's head, saying, 'There, you came for salt, now take some.'"[13]

Though deterred by terrain and stout Confederate defenses in October 1864, Union officers never abandoned plans to destroy the South's last major supply of sodium chloride. In December, Union soldiers returned, this time under the command of Gen. George Stoneman, with Burbridge commanding a division. With the war going badly for the Confederacy, some of those who had previously defended the town had been deployed elsewhere. The men who remained again took to the high ridges, but this time Stoneman's soldiers overwhelmed Confederate forces in less than twenty-four hours. After being held off by fire from two forts during the afternoon, some of Stoneman's units dismounted and advanced uphill under the cover of darkness to the edge of the forts, before launching an attack that caused the defenders to panic and run. The Union soldiers then did their best to destroy the furnaces and kettles that had provided salt to the Confederacy for nearly four years. Stoneman's troops were only partially successful. Almost as soon as the Union men moved on, local people partially rebuilt the works and by March 1865 salt production had resumed, albeit on a much smaller scale.[14]

For much of the war, topography—what commanders saw as "the lay of the land"—usually proved more important to military strategy than salt or any other mineral buried beneath. As at Saltville, the combination of limestone and more erosion-resistant rocks loomed large in some of the war's most decisive campaigns. In the spring of 1862, the Great Appalachian Valley, a long level corridor underlain by limestone, proved vital to Confederate strategy. As Gen. George B. McClellan approached Richmond from the east with a large army, Gen. Robert E. Lee ordered Stonewall Jackson to create havoc in the Shenandoah Valley in western Virginia to prevent any more Union troops from reinforcing McClellan's army. Beginning in May 1862, Jackson used the ridges as blinds and the valleys as highways to move rapidly and befuddle parts of three Union armies. Over the course of one month, he kept approximately 50,000 Union troops occupied, preventing any substantial reinforcements from reaching McClellan. His actions made it possible for Lee to attack McClellan and drive him away from Richmond, saving the Confederate capital.[15]

In the late summer of 1862, Lee again used the Great Appalachian Valley to move his army out of Virginia to Sharpsburg, Maryland, and the region along Antietam Creek. The Battle of Antietam took place in the northern reaches of the Great Valley, where uneven erosion of sandstone, dolomite, and limestone had created a landscape of shallow depressions, hollows, and rocky outcroppings. David Miller's cornfield, site of the horribly bloody encounter during the early hours of the battle, lay on thirty acres of limestone-enriched farmland. Later, 2,200 Confederate troops gathered behind a breastwork of fence rails in the Sunken Road, a ditch-like thoroughfare created by limestone erosion and heavy wagon traffic. From there, the southerners held off 9,000 Union men until midafternoon, when northern forces made their way to a higher ridge, underlain by erosion-resistant rock, from which they could fire down onto the southern line. By the time Confederates gave up the fight, 5,000 more men lay dead or wounded in or near what would thereafter be known as "Bloody Lane." Similar trends emerged during the final phase of the battle, during which 500 Confederates, well positioned on high ground, effectively delayed Gen. Ambrose Burnside's crossing of Antietam Creek over a bridge that now bears his name. Union soldiers noted, "A slight rise of ground in an open field, not noticeable a thousand yards away, becomes in the keep of a stubborn regiment, a powerful headland against which the waves of battle roll and break."[16]

Eighty miles beyond the westernmost ridges of the Appalachians, Stones River sliced through similar limestone deposits in the Tennessee Piedmont. There, on December 31, 1862, the Union army of the Cumberland found itself pressed nearly to the breaking point. As southern troops moved to cut off the Federals' planned escape via the Nashville Turnpike, Gen. William Rosecrans desperately needed to delay the Confederate assault long enough to shore up his defenses. Rosecrans's troops made their stand amid a peculiar formation of limestone and shale near the center of their line. There, the limestone rocks had eroded from around the shale, leaving a maze of trenches between larger rocks and outcroppings. Thin soils made the area ill-suited for agricultural clearing, and dense cedar forests, well adapted to limestone soils, flourished in and around the odd formations. Together the rocks and woodlands provided nearly ideal cover for Union troops. One Union soldier noted that combatants found "the cedar woods so thick, and so filled with rock and caverns and fallen trees, that it was almost impossible to get through it."[17]

The Union men held off several Confederate advances, providing Rosecrans with the time he needed to establish a stronger defensive line. Both sides paid a fearful price in dead and wounded. A Confederate officer recalled that during the attacks, "men fell around on every side like autumn leaves and every foot of soil over which we passed seemed dyed with the life blood" of his soldiers. Northerners named the geologic formation "the Slaughterpen," because the lifeless bodies and pools of blood left behind reminded them of the Chicago stockyards. Two days later, northern forces moved onto the higher ground along the river's only available ford. From there they rained artillery fire on southern troops, inflicting some 1,800 casualties. The Union suffered heavy losses as well but held Middle Tennessee, prompting Lincoln to remark that Rosecrans had achieved "a hard-earned victory, which had there been a defeat instead, the nation could scarcely have lived over."[18]

Six months later, topography and terrain helped determine the course of the war's most famous battle. Using the Blue Ridge Mountains to conceal his troops, Robert E. Lee moved into Pennsylvania, hoping to capture the state capital of Harrisburg and achieve a key victory that might end the war. Learning that the Union Army of the Potomac was rapidly closing on him, Lee opted to reunite his army at a central location in the southern part of the state: Gettysburg. Geologically speaking, the town lay in a giant basin atop limestone and shale formations. The level terrain around Gettysburg was such a prominent feature that it had become a junction for no fewer than ten roads that dissected the surrounding ridges. Lee's men marched down the roads from the west and north, while the Army of the Potomac came from the south and southeast.[19]

Though scholars continue to debate the strategies and decisions surrounding the Battle of Gettysburg, the deadly three-day encounter essentially evolved into a struggle for the higher ground around the level basin. The Army of the Potomac's tenacious defense of the rocky outcrops of Little Round Top and Culp's Hill thwarted Lee's plan to overwhelm each of the Union flanks on July 2. On the southernmost flank, early in the afternoon of July 2 Union engineering officer Gouverneur Warren climbed to the top of

Little Round Top, which rose 150 feet above the surrounding valley, and observed Confederate brigades getting into position to attack the undefended hill. Recognizing its importance to securing the Union position, Warren immediately ordered two brigades to the rocky slopes just in time to hold off a determined Confederate assault that cost enormous casualties on both sides.[20]

On the northern flank, Union troops constructed breastworks on Culp's Hill, rising 160 feet above the Rock Creek valley, which allowed one Union brigade of about 1,400 men to hold off an entire Confederate division three times its size on the evening of July 2. Confederate troops quickly became fatigued as they struggled up the slopes, adding to the Union's defensive advantage. Lee's doomed frontal assault on July 3 — now immortalized as "Pickett's Charge" — required 13,000 men to cross three-quarters of a mile of undulating fields to attack Cemetery Ridge. Though the ridge only rises forty feet above the surrounding fields, the long approach over open terrain left Lee's men exposed to destructive artillery and rifle fire. The attack failed spectacularly, costing 50 percent of the attacking force and essentially spelling the end of the battle.[21]

The prevalence of limestone formations at major battle sites has not been lost on geologists interested in the Civil War. In surveying rock formations associated with the conflict's twenty-five bloodiest campaigns, military geologists Bob Whisonant and Judy Ehlen note that a quarter of those battles — all of them engagements with enormous casualty counts — occurred on ground underlain by limestone. Though the scientists speak facetiously of "killer limestone," they are quick to note that it was limestone's propensity to erode faster than the surrounding terrain that created the uneven ground on which so many soldiers died. Whether topography proved an offensive or defensive advantage also depended on a host of other factors, including previous human use. Presented with suitable limestone-enriched soils, local people planted corn and wheat. Where soils were thin, they left woods and rocks intact. Where streams cut deeply into bedrock, they built bridges or resorted to available fords. All of these actions proved crucial when the armies came. Though command decisions by Union and Confederate officers mattered, the physical landscape — the product of long-term interaction between people and nature — usually dictated the strategy and pace of fighting.[22]

Nowhere was that blending of human and natural forces more apparent than during a battle fought in May 1864 on a large tract of once-forested land in Spotsylvania County, Virginia. Soils there were typical of the southern

Piedmont: sandy loams and clays created by millennia of mountain erosion, fairly well drained, and naturally acidic. Generally speaking, the ground gave rise to hardwood forests consisting primarily of various oaks and hickories. Long before colonization by Europeans, Virginia's Native people lit ground fires to facilitate agriculture, hunting, and travel. As a result, those trees that resisted fire, like red oak, became more prominent in the region. European colonization brought more intensive cultivation of soil-depleting crops such as tobacco and corn. Colonists initially used Indian methods of land clearing such as girdling larger trees and burning underbrush, but, over time, fields became more open, with little natural vegetation to break up or scatter rainfall. In the hilly Piedmont, heavy precipitation brought more extensive erosion that further decreased fertility. By the early 1730s, many of the region's former tobacco lands had been abandoned. Small pines and scrubby oaks took root in the acidic soil, prompting travelers to describe the tracks as "poisoned lands" or "poisoned fields." About that same time, some Virginians began to refer to the area by a name now well known to Civil War historians: "the Wilderness."[23]

As prospects for tobacco waned, Virginians turned their attention toward another potentially valuable resource deep within the earth: iron. Off and on from the mid-1730s until the 1840s, various individuals and small corporations operated furnaces capable of extracting iron from small deposits of hematite, called "black ore." Easternmost of these operations, the Catharine Furnace, built in the winter of 1837–38, stood fifteen miles west of Fredericksburg in Spotsylvania County. For the most part, fuel for the Catharine Furnace came from charcoal, essentially a form of porous carbon, created by slow burning of green timber. During the ten years the furnace operated, workers cut nine to ten square miles of surrounding forests, some of which had sprouted on old fields abandoned a century before during the tobacco boom. When operations ceased at the Catharine Furnace in 1847, the cutover tracks began the slow, haphazard process of reverting to forest.[24]

Generally speaking, thoroughly cleared lands in the Virginia Piedmont first sprout various grasses and weeds, followed in short order by cedars and pines. Hardwoods such as sweet gum (if the area is wet enough) and tulip poplar follow, eventually giving way to oaks and hickories. In this particular part of Spotsylvania County, however, colliers in search of the best wood for charcoal had selectively cut oaks, leaving behind stumps and root systems that remained alive. Sprouts, commonly known as suckers, quickly erupted from the stumps. The new shoots drew on energy reserves in the roots, sometimes reaching a height of ten feet in only ten years. In addition to the various pines, cedars, and other species that initially favored open ground, the sprouting stumps helped create a thick tangle of saplings, shoots, vines, and weeds. By 1864, having grown for seventeen years since the iron furnace shut down, the woodlands, lacking tall or thick trees, probably resembled the "poisoned lands" left behind by colonial tobacco farmers in the 1730s. Though local people still called it the Wilderness, it had, in fact, been shaped by human activity for more than two centuries.[25]

On May 4, 1864, Ulysses S. Grant moved the Army of the Potomac into the region, hoping to meet Lee's army in the open fields south of the dense, shrubby forests. As three Union corps moved south through the Wilderness on May 5, however, Lee attacked from the west with two of his three corps. James Longstreet's corps had farther to travel and would not arrive until the next day. As soon as Grant became aware of Lee's approaching army, he turned to engage along the two main roads through the dense woods—the Orange Turnpike and the Orange Plank Road. Fighting was chaotic and visibility limited. Of the fight along the Orange Turnpike, a Confederate soldier wrote, "A more difficult and disagreeable field of battle . . . could not well be imagined."[26]

The fighting raged fiercest over the few cleared acres of small farm fields in the otherwise impenetrable thickets. A Union soldier remarked that his unit had to move through "woods where one thinks that not even a hare can get through." Neither side could clearly see their opponent. One Yankee declared it "a weird, uncanny contest—a battle of invisibles with invisibles." The woods soon blazed with fire, turning the scene into an inferno that swallowed up wounded soldiers. A Union soldier recalled, "The incessant roar of the rifle; the screaming bullets; the forest on fire; men cheering, groaning, yelling, swearing, and praying! All this created an experience in the minds of the survivors that we can never forget."[27]

About four hours after forces clashed on the Orange Turnpike, Union troops attacked Confederates three miles south on the Orange Plank Road, in an equally confused and bloody attack in the woods. One Confederate recalled the fight as "butchery pure and simple. . . . It was a mere slugging match in a dense thicket of small growth, where men but a few yards apart fired through the brushwood for hours, ceasing only when exhaustion and night commanded a rest." A Union division sent to exploit the gap between the widely separated Confederate forces got hopelessly entangled in the trackless forest and rendered no assistance. Only with darkness did the fighting subside. Neither side had gained or given up substantial ground, a stale-

The thick, shrubby woods of the Wilderness shortly after the battle.
Courtesy Library of Congress.

mate that insured heavy casualties. Before dawn the next morning, Grant amassed troops to attack Lee's right along the Orange Plank Road. The fate of Lee's army, and the outcome of the war, hung in the balance.[28]

At dawn on May 6, a Federal attack overwhelmed the right wing of Lee's army, but Longstreet's corps arrived just in time from the west, providing critical reinforcements that, for the moment, helped shift the course of the battle. Longstreet's scouts soon discovered jutting through the woods an unfinished railroad cut—land that had been graded and cleared but never laid with tracks. It allowed for concealed movement around the Union left flank. Longstreet sent four brigades on the path. They launched a surprise attack just before noon that rolled up the Union flank and forced it into a hasty retreat. The Confederates, it seemed, might win a crucial victory. Longstreet and some of his aides immediately rode forward on the Orange Plank Road to determine how best to press their advantage, but nature, shaped by earlier human activity, intervened.[29]

As they ventured on their scouting expedition, shrouded in smoke and the deepening shadows of the foreboding woods, their own units mistook them for the enemy and struck them down with a volley, grievously wounding Longstreet—an event eerily similar to Stonewall Jackson's mortal wounding nearby almost exactly a year earlier. Lee's most trusted subordinate would be out of action for five months. Longstreet's wound stopped the Confederate momentum, and a renewed assault several hours later failed to dislodge the entrenched Union troops. At nearly every turn, the uniquely forested landscape, the fires, and poor visibility profoundly affected the course of the battle.[30]

In one sense, the Wilderness and other major engagements illustrate what military historians have long understood—namely, that armies in the field must either turn the surrounding landscape to their advantage or find ways to overcome it. What scholars have sometimes lacked, though, is an understanding of terrain as more than a static geographic stage for human action. Like everything in the natural world, battlefields were living entities subject to constant change. Some of nature's alterations, including the effects of geologic anomalies and erosion of limestone, occurred so slowly that human minds can barely fathom the process. Changes wrought by people, notably deforestation resulting from agriculture and the demand for wood to fire salt and iron furnaces, occurred rapidly and reflected the shifting rhythms of economies, technologies, and human values. Anywhere Union and Confederate troops met on the battlefield, they did so on terrain that had an extended history, one shaped both by nature and people long before the generals and soldiers arrived on the scene. Unraveling that long-term interaction between people and nature improves our understanding of terrain as the context for war. Perhaps more important and unexpected, it can also illustrate the ways that war changed terrain, changes that endured for decades to come. One of the best examples of long-term change engendered by the Civil War comes not from the countryside but from one of the South's most prominent cities: Atlanta.

In 1837, surveyors for the Western and Atlantic railroad carefully sited the location of their terminus on elevated ground six miles south of Georgia's Chattahoochee River. From there, the city of Atlanta grew in a circular pattern, six miles in diameter, with the train depot at the center. By 1861, forty-four trains a day passed into and out of a city that by then had nearly 10,000 residents. During the war, as Atlanta's military importance increased, its population grew to more than 20,000. With housing for newcomers scarce, tent villages sprang up around the city. As was often the case on rural battlefields, the sudden influx of people created myriad environmental problems. Water had always been in relatively short supply in Atlanta. In 1859, an insurance adjuster classified Atlanta's water supply as "deficient," not least because the city's shallow wells and water reservoirs had sometimes proved incapable of dousing major fires. By 1862, the city's privies could no longer handle the massive influx of human waste, creating untold sanitation problems. The arrival of Sherman's army in July 1864 pushed Atlanta to the verge of ecological collapse.[31]

Shortly after noon on July 20, 1864, as Union forces moved to within two miles of downtown, the first shells from their twenty-pounder Parrott rifles fell on the city. A five-week bombardment followed, designed to force evacuation of the Confederate army defending Atlanta. Sherman vowed to make the city "too hot to be endured." He was as good as his word. One civilian decried the apparent inhumanity claiming, "This seems to me to be a very barbarous mode of carrying on war, throwing shells among women and children." The bombardment killed about a dozen civilians, wounded many more, wrecked much of the city's infrastructure, and left survivors with frayed nerves and little hope for the future. By the time the shelling ended on August 25, Union artillery had fired more than 100,000 shells into the city—an average of nearly 2,800 a day. As one journalist explained, "All the fires of hell, and all the thunders of the universe seemed to be blazing and roaring over Atlanta." The shells ripped apart the city's dwellings. "Such houses!" an Illinois soldier wrote upon viewing them after the battle. "Each one riddled and torn by our shells, here a tall chimney knocked down and there a portico carried away." With the water supply even more restricted due to unrelenting summer heat and drought, fires burned through some 90 percent of the 3,600 homes in the center of Atlanta. When Sherman severed the last railroad supplying the city on September 1, the Confederate defenders abandoned the city.[32]

Sherman's army took control of Atlanta on September 2, and just over a week later, he ordered the remaining 4,000 residents removed from the city and its environs. On the heels of the bombardment, the mandatory evacuation order brought protest from city officials. Atlanta mayor James M. Calhoun and two city councilmen wrote a petition to General Sherman on September 11, pleading that civilians not be forced "to wander strangers and outcasts, and exiles, and to subsist on charity." Sherman was unmoved, responding the next day, "War is cruelty, and you cannot refine it; and those who brought war into our country deserve all the curses and maledictions a people can pour out." Sherman declared that the forlorn residents would have to fend as well as they could "until the mad passion[s] of men cool down, and allow the Union and peace once more to settle over your old homes at Atlanta." Between November 10 and 15, Sherman's army burned much of the city as it departed and began its famous "March to the Sea." Any buildings and trees that had survived the siege went up in flames.[33]

Scholars have rightly focused on the ecological havoc Sherman wrought as he made his way from Atlanta to Savannah. Aided by an atypical stretch of dry weather, the general made good on his promise to "make Georgia howl," as he put the torch to agricultural fields and butchered Confederate livestock to feed his troops. According to environmental historian Lisa M. Brady, "In targeting the South's agricultural sector, the Union strategy undermined the region's most basic relationship to the natural world, destroyed the Confederacy's ecological foundations, and assured federal victory."[34]

Even so, the environmental chaos Sherman triggered in Atlanta probably had a more enduring legacy, one that historians are still struggling to understand. The bombardment destroyed many of the city's already inadequate wells and cisterns. The siege also wrecked all but one of Atlanta's water pumps. Some residents claimed that Union soldiers intentionally contaminated other water sources with salt. After Sherman left, city physicians and chemists declared water from the center city's remaining wells unfit for consumption. As peace returned, those who sought to rebuild Atlanta faced the ongoing problem of providing water for an urban center with an increasing population. By 1880, the city had more than 37,000 residents, nearly four times as many as in 1860.[35]

As in other southern cities, civic officials made sure that the rebuilding of Atlanta reflected their segregationist politics. In Atlanta, those policies had distinctly environmental overtones. Wealthy whites built their new residences on the higher ground in the city center and north along Peachtree Street. To accommodate the new affluent neighborhoods, the city channeled sewage out of downtown onto the surrounding low-lying terrain. Barred from the all-white districts, freedpeople who migrated to the city inevitably had to settle near streams and rivers on the outskirts of town where waste from the high ground accumulated. When it rained, the marshy land flooded and soon became awash in sewage. Diseases from polluted drinking water ravaged black neighborhoods. Even whites in the city center could sometimes smell the stench of the open ditches and had "to sleep with their windows down, even on hot nights." As the population expanded and the high ground of the central city filled up, wealthier whites moved out to other elevated regions, leaving black residents to fend for themselves in the polluted lowlands. With names like Druid Hills and Ponce de Leon Springs, the suburbs promised healthier land and sparkling water, resources that had been in short supply since 1864. In many ways, Sherman's siege set in motion patterns of segregation and environmental racism that plagued Atlanta for decades to come.[36]

The activities of encamped armies compounded the city's water problems. Union and Confederate troops cut trees in and around the city for fortifications and fuel. One Union soldier noted, "The handsome streets are dug up in crazy places for fortifications; fine shade trees are hacked to pieces." Exploding ordnance also took a heavy toll. Estimates suggest that more than two-thirds of the trees within the city limits perished in the siege. As one Union officer explained, "Many of the ornamental trees are cut down by our shells." According to another witness, "Shade trees a foot through are cut off" by the shelling. Without vegetation to soak up rain- and wastewater, it took years for the city to rectify problems created by increased erosion and runoff. Visitors to Atlanta in the decade after the war frequently remarked on the city's absence of trees, while local newspaper editors implored residents to plant trees well into the 1900s. One observer for *Harper's Magazine* noted in 1879 that earthworks close to the city were still visible and barren of ground cover. The soil, he wrote, "takes so slowly to new planting that for fifteen years compassionate Nature has tried in vain to hide their marks under her mantle of herbage and wild shrubbery." It took the better part of a century after Sherman's siege, and express efforts to replant, for Atlanta to regain even a fraction of the tree cover it had enjoyed before the two armies dug in around the city in August 1864.[37]

Atlanta's experience hardly proved unique. Over the course of the war, commanders on both sides ordered soldiers to establish ever more elaborate trenches and other defensive works, all of which required wood for construction and reinforcement. In addition to those needs, local timber furnished fuel for the ubiquitous campfires over which enlisted men cooked their food and warmed their bodies. Even in the temperate South, winter brought increased demand for firewood and timber to build winter quarters — usually crude cabins with chimneys and plank floors. A host of other activities crucial to military operations also required lots of wood. Under the watchful eye

Confederate fortifications northwest of Atlanta, near Peachtree Street. The defenders stripped the surrounding landscape of trees to build their defenses and provide clear fields of fire. Courtesy Library of Congress.

of army engineers, soldiers laid thousands of logs across muddy or swampy terrain to create corduroy roads that facilitated movement of troops and armaments. Each mile of railroad track required 3,000 ties, or "sleepers," between the rails. By 1865, the repair of damaged railroads caused forests to vanish at an astounding rate. Millions of hemlock, chestnut, and oak trees went into the construction of railroads, which used nearly 6.5 million cords of fuel annually. Hardwoods in the woodlands on the farms along the tracks nearly vanished.[38]

Armies desperate for wood first looked to the most convenient source: the fences and other structures on nearby farms. Virginia rail fences, the most common enclosures in the South, consisted of interlocking sets of six to ten rails that zigzagged along agricultural fields to keep out livestock. Every mile of fencing required 6,000 to 7,000 such rails and could provide an army with wood that had already been split and seasoned for several years. Abandoned barns and other structures could also be torn apart for firewood and building material. In 1863, a Union surgeon in Culpeper, Virginia, explained, "An unoccupied house or barn dissolves in an hour. One after another takes a board, and it is soon gone." In January 1865, one of Sherman's soldiers remarked upon a South Carolina campsite that had "fine houses" nearby. "As soon as we stacked arms," he wrote, "the whole brigade rushed hell upon these houses and in half an hour not a vestige of them remained save the chimneys."[39]

When the fences and farm buildings disappeared, soldiers turned their axes on the forests. Pines, perhaps the most common trees on the southern landscape, burned quickly and produced clouds of acrid, sooty smoke. Troops much preferred cedar, hickory, and oak, which burned more slowly and provided steady heat over longer periods. Those trees, along with hemlock from the southern uplands, also provided the best wood for buildings and road construction. It did not take long for hardwoods to grow scarce around the encampments, especially if armies remained for extended periods. Near Fredericksburg, Virginia, Confederate soldier Murdoch McSween wrote in the winter of 1863 that Lee's army quickly denuded the landscape of its prized hardwoods, "the timbered tracts [that] have long been more highly valued and carefully preserved than the open land." Waxing eloquently about lost nature, he lamented, "Very soon not a vestige of the regional forests will remain, and those majestic old oaks, that witnessed the spirits and vows of the red man, and afforded comfort and protection to the banished and persecuted of our own race, can never again shelter either exile or invader."[40]

In March 1863, a Union soldier stationed across the Rappahannock River from Fredericksburg marveled at "how the whole country round here is literally stripped of its timber. Woods which, when we came here [in November 1862], were so thick that we could not get through them any way are now entirely cleared — the pine being used for building and making roads, and the cedar and hard wood, of which there is a great quantity, for fire wood." Soldiers seeking warmth did not even leave stumps behind in their search for kindling. One Connecticut soldier wrote in April 1863, "We are now gathering what we call the second crop, that is, we cut off the stumps even with the

Winter camp of the Sixth New York Artillery in Virginia, April 1864. Soldiers used the nearby timber to build wood cabins and corduroy a sidewalk to keep themselves out of the mud. Courtesy Library of Congress.

ground." He noted that when residents returned to the region, "they won't find a piece of timber large enough to make a respectable souvenir." Writing that same year, a Pennsylvania soldier proclaimed, "You have no idea what an army can do—it is worse than seven year locusts."[41]

When soldiers left camp to engage the enemy, forests near the battlefield felt the ravages of war. Exploding shells destroyed trees; continuous volleys of bullets stripped away leaves and mangled bark. At Gettysburg, a Union surgeon who walked the ground at Culp's Hill after the battle found that "the trees were completely cut to pieces by bullets." In one tree he counted 150 projectiles embedded into the bottom eight feet of the trunk. On the field at Chickamauga, an Illinois soldier marveled at the terrible evidence of human destruction he saw in the battle's aftermath. "Hundreds of large trees were cut completely into with shot, and hundreds of items of immense size are bored through and through with those iron messengers," he wrote, "while the butts of them were so riddled with bullets that in many of them not a space of a single square inch can be found free from the scar of a bullet." The millions of bullets and artillery shells fired in battle left another legacy in the shredded forests. Over time, the spent projectiles leached lead into the soil. Certain lead oxides might remain present in the ground for thousands of years, wiping out bacteria and fungi that helped facilitate decomposition and regeneration of topsoil. In some cases, the contaminants interfered with photosynthesis in plants that moved in after the battles.[42]

The exact extent of deforestation is difficult to determine. Historian Megan Kate Nelson notes that if one campfire per year consumed between ten and twenty acres of woodland (as some studies suggest), then the armies "consumed at least 400,000 acres of trees for firewood alone" each year. She further estimates that it took 60,000 board feet of timber (equivalent to 200 mature trees, or a little more than 3 acres) to corduroy one mile of road. Her calculations indicate that the armies probably cut 1.6 million acres of trees for firewood and another 30,000 acres for roads. Wood to power steam locomotives took an even heavier toll, perhaps accounting for as much as a half million acres of timber *per year* for the duration of the war. Considering wood needed for shelter, fortifications, and other miscellaneous purposes such as salt furnaces, refineries, and smelting, a conservative estimate indicates that about 4 million acres of trees fell in support of the war.[43]

Wood shortages made life tough on civilians. Farmers often lacked suitable rails to rebuild fences around their crops. Open range livestock, which normally sustained themselves in the forests, now ravaged unprotected fields. In the long term, deforestation reduced the soil's ability to absorb and hold moisture, resist erosion, and regrow natural vegetation. According to historian Erin Stewart Mauldin, deforested land became "less suited to intensive cultivation during a time when so many farmers pursued cash crop agriculture." The destruction of timber might have forced some civilians to relocate. In 1865, a U.S. Department of Agriculture agent heard rumors that Virginia refugees were not returning home "because the war has swept away the trees."[44]

Such ecological effects often went unnoticed by most observers. Civilians only knew that, after the war, it became much harder to make a living in regions where armies had camped for long periods. In Virginia, Spotsylvania County was home to four major battles, and occupied by large armies in the winter of 1862–63 and 1863–64. Henrico County (of which Richmond is the county seat) also experienced the ravages of battle, siege, and long encampments in 1862 and for the final ten months of the war. Between 1860 and 1870, Spotsylvania County lost 44 percent of its improved acreage. Wheat production dropped by 58 percent, corn production fell 59 percent, and tobacco production declined by 92 percent, while Henrico lost 10 percent of its improved acres, 62 percent of its wheat production, 64 percent of its corn, and 98 percent of its tobacco. In contrast, Pittsylvania County (of which Danville is the seat) in southern Virginia, and Amherst County in west central Virginia, just north of Lynchburg, hosted no armies or battles. Pittsylvania's improved acreage was only down 3 percent in 1870. Its wheat production fell 32 percent, corn declined 41 percent, and tobacco dropped by 39 percent. Amherst County gained nearly 17 percent in improved acres, while it lost 28 percent of its wheat, 49 percent of its corn, and 55 percent of its tobacco by 1870.[45]

Losses in the counties of Spotsylvania and Henrico were far more severe, illustrating the devastating and lengthy effects of battles. With fence rails gone, remaining livestock ate crops in the fields. The armies had cleared the woodlands for firewood and winter cabins, so farmers could not replace their fences or repair their structures. With draft animals in short supply, residents used fewer plows and transport wagons. In Spotsylvania County, farmers were unable to till even 60 percent of the land they had farmed before the war. Farmers in both counties nearly abandoned tobacco altogether, though almost none had switched to cotton farming by 1870. They invested almost exclusively in food crops, and still they produced less than half as many comestibles as they had in 1860.[46]

In terms of lasting effects on southern terrain, the armies' need for timber probably had less impact than the demand for one of the small pleasures of war: tobacco. As Bell Irvin Wiley once noted, "It is doubtful if any single item except food, water, and letters from home was so highly cherished by Johnny Reb as 'the delightful weed.'" It quickly became so essential to troop morale that southern generals monitored the tobacco supply almost as closely as enemy troop movements. Fears of shortages multiplied as supplies from northern and eastern Virginia, Maryland, and Kentucky diminished in the wake of fierce fighting and Union occupation. Enterprising farmers in southern Virginia and North Carolina rushed to fill the void. Even so, tobacco processors had to seek safer ground away from the fighting. Several large Richmond tobacconists moved their operations to the more peaceful confines of Danville.[47]

With southern armies starving in the field, the Confederate government constantly urged citizens to plant more food crops, especially during the latter stages of the war. According to historian Drew A. Swanson, many

Piedmont farmers defied such orders, favoring profits over patriotism. One soldier insisted that his wife continue planting tobacco instead of foodstuffs on their Pittsylvania County farm, because, he argued, "I had rather had [tobacco] than Confederate bonds." Another farmer suggested that southerners should "make all the tobacco we can" because "the article is bound to sell high." In the autumn of 1863, a North Carolina newspaper editor excoriated the greed and lack of commitment among such Piedmont farmers: "It would be a glorious deed for this Southern Confederacy if every Tobacco Factory in it were burnt to the ground and their very ashes scattered to the four winds of heaven." He averred, "The people can do better without tobacco than meat and bread." Such appeals did little to alter planting practices.[48]

Changes in the southern labor system after the war also had long-term environmental consequences. Historians are only beginning to understand how the emancipation of slaves might have affected the landscape. Before the war, the southern practices of shifting cultivation meant that farmers kept roughly one-third of their land in crops. The rest consisted of forested tracts for future clearing or old fields lying fallow to recover fertility. Though ecologically efficient, as long as one had adequate land and slaves, the system proved incredibly labor intensive. In addition to clearing woodlands for new fields, slaves had to ditch around older agricultural tracts to prevent flooding and erosion. Crops had to be fenced to keep out livestock and manure collected to fertilize planted fields. After the war, as contract labor replaced slave labor, such practices became less common. As a result, Erin Stewart Mauldin argues, "the decreasing frequency of land maintenance exacerbated erosion, soil nutrient deficiencies, and crops' susceptibility to drought and damage by livestock."[49]

It did not take long for careful observers to notice the change. A Freedmen's Bureau official in Marengo County, Alabama, explained it this way in 1868: "Men who are free will not clear new land . . . while the breadth of land annually thrown out of cultivation is about the same as formerly." The refusal to ditch increased flooding and allowed topsoil to wash into creeks and rivers. Even increased use of commercial fertilizers could not replace nutrients lost from erosion of topsoil. One expert suggested that soil erosion in the Carolinas and Georgia rose between 70 and 120 percent between 1860 and 1880. These factors led to declining yields in pounds of cotton and bushels of corn per acre. The war ended slavery, but the new labor system, in conjunction with southerners' dependence on cash crops, brought a new round of environmental degradation.[50]

In time, the changes wrought by postwar shifts in agriculture and labor became less noticeable. One effect of the war on terrain, though, remains visible to this day. Thanks to the National Park Service and other agencies, some seventy Civil War battlefields have been preserved and opened to the public. Much like uniforms, firearms, ammunition, swords, and flags, battlefields offer visitors a chance to experience what historian Michael DeGruccio calls "the symbolic and spiritual power of things." For many visitors that tactile experience—walking over the same ground as Union and Confederate troops, seeing what they saw, touching what they touched—makes the war seem less distant.[51]

Nowhere are the effects of the war on modern terrain more visible than on the outskirts of Richmond. There, amid big box stores, strip malls, and residential neighborhoods, small patches of woods and open fields stand undeveloped—tiny unoccupied islands in a sea of suburban sprawl. Modern society has not yet swallowed up these areas for one reason: they mark the sites of some of the Civil War's most intense battles, including those of the Seven Days of 1862. At Beaver Dam Creek, a handful of acres preserving a small slice of the battlefield allows visitors to get a sense of the swampy terrain near the Chickahominy River that often stymied both armies, and proved calamitous for Confederate forces attacking that position. At Gaines Mill, a walking trail winds down from a plateau through thick woods along a swampy creek, offering visitors a glimpse of the difficulty of assaulting such a naturally imposing position.

Malvern Hill is perhaps the most moving of the sites near Richmond. The National Park Service owns nearly 1,000 acres of fields and woodlands that preserve the landscape of the battlefield much as it looked in 1862. Visitors can walk the perimeter of a large open field through which Confederates attacked Union forces and artillery situated on the higher ground. The woods around the site once provided cover for anxious troops waiting to attack and wounded soldiers who somehow escaped the carnage. To stand at the perimeter of those woods, looking across that open terrain and knowing what awaited those going up the hill is proof positive of how valuable high ground could be, especially in the swampy lowlands of the peninsula. Walking the 500 yards from the Confederate position to the Union batteries, imagining the thousands of artillery shells and rifle bullets filling the air every step of the way, is a sobering experience.[52]

Fifty miles north of Richmond, at the site commemorating the Battle of Fredericksburg, the terrain can evoke similar feelings, but there the modern world intrudes more directly on the experience. For example, the famed sunken road at the foot of Marye's Heights once held Confederate troops who could gaze across 600 yards of undulating fields toward the town from which seven divisions of Union troops launched fourteen futile attacks. None got closer than forty yards. While the road remains behind the battlefield's visitor center and bookstore, the view does not. Visitors today can only see the neighborhoods and urban development that have swallowed the battlefield. Similarly, visitors standing atop Lee's Hill no longer get the same panoramic view of the Rappahannock River that the Army of Northern Virginia's commander had as he directed the battle on December 13, 1862. At the war's centennial, they could. Shortly before the 1862 battle and again in 1963, the hillside timber was cut, providing a clear view for both Lee and twentieth-century visitors. However, a modern industrial park grew up in the valley between the hill and the river, and the National Park Service has chosen to allow nature to grow unchecked. Tall trees now shelter visitors from the sight and noise pollution of the industrial park. The Park Service only allows a narrow viewing corridor in the direction of Marye's Heights to remain.[53]

Fifteen miles west of Fredericksburg, visitors who seek out the Wilderness battlefield may think they are seeing the same natural setting that greeted the soldiers in May 1864, but they are mistaken. At the time of the battle, the landscape was a thick cluster of scrub forest that had grown unmanaged for seventeen years. As a result, practically no tree was taller than twenty feet, and few were thicker than six inches in diameter. The Park Service works hard to maintain the original sight lines of the battle, but most of the surrounding forests are now populated by giant hardwood trees, several of which stand 100 feet tall and thirty inches thick. The roughly 180,000 soldiers who fought in these woods would have considered it a blessing to have such trees to hide behind when the lead was flying during that grim battle. Instead, they had to fire on nearly invisible enemies through entangled thickets of briars and vines. Maneuvering through the shrubby vegetation was a skin-scarring, fabric-ripping, frustrating task that few modern Americans could abide. Visitors can at least gain some sense of what the thick shrubby terrain was like by walking a carefully maintained trail through a copse of younger trees, near the intersection of the Orange Plank and Brock Roads. The trail winds through a dense cluster of saplings and older trees, but nowhere are tourists completely removed from the sounds of traffic. The National Park Service rangers do a thorough job of interpreting the battle at its exhibit shelters, but they cannot trap the 1864 landscape in time.

Sites of the most famous Civil War battles have gotten more attention but nonetheless face similar problems. The site of the war's single bloodiest day, Antietam, provides a case in point. At first glance, the modern Antietam landscape might be easily mistaken for the one of 1862. Antietam Creek still flows swiftly through its small valley cut in the native limestone. During the warm months, corn grows in the same limestone soils that once nourished David Miller's crop. From almost any vantage point, visitors can take in the rolling terrain and remnants of the ragged forests that concealed troop movements. The Sunken Road remains clearly visible, allowing those who walk it to imagine the carnage that turned it into Bloody Lane. A carefully restored Burnside's Bridge can be found exactly where it stood in 1862. The land's pastoral character clearly reflects the mid-nineteenth-century obsession with neatness, order, and productivity that defined northern agriculture on the eve of the war.

Yet all is not quite as it seems. Preserving the battlefield has been no easy task. The long and difficult process of land acquisition began even before the war ended. In 1864, the Maryland legislature launched efforts to buy a tract east of Sharpsburg for a state and national cemetery to honor the Union dead, a purchase completed in 1865. The high cost of maintenance eventually led Congress to take over administration in 1877. In 1890, after much lobbying by various Civil War veterans groups, Congress expanded its holdings and jurisdiction to include a tiny parcel of the original battlefield. Other piecemeal acquisitions followed, but the battlefield as preserved today is much smaller than in 1862.

Some of those involved in the early phases of Antietam's preservation declared, "The field on which the battle took place is practically unchanged from what it was on the day of the action, save the cutting down of some trees, and presents to-day, as it did in 1862, the most open field on which was fought any of the great battles of the rebellion." But as with other such landscapes, Antietam is subject to the whims of both people and nature. Much like those who occupied the landscape just before the battle, National Park Service employees devote considerable time and money to preserving Antietam's neat 1862-like appearance. Some modern practices closely resemble those of the nineteenth century. The Park Service periodically burns fields to promote growth of certain grasses and harvest crops. Workers maintain orchards and thin forests. Maintenance crews cut back encroaching woodlands and try to prolong the lives of "witness trees," so named because they were there in September 1862. At the same time, today's managers have to eradicate invasive species, control various wildlife populations, and mediate the effects of water and air pollution.[54]

To an extent, the same is true of Gettysburg. In addition to keeping up the landscape with its fields and orchards, the Park Service and private organizations go to great lengths to preserve original sight lines from 1863. In 1992, media mogul Ted Turner—fresh off his bit part in the movie *Gettysburg*—donated $50,000 to have the power lines along the Emmitsburg Road buried. Visitors at the "Angle" on Cemetery Hill no longer see those anachronistic signs of modernity as they gaze across to Seminary Ridge and imagine 13,000 Confederate troops marching toward them in Gen. George Pickett's ill-fated charge on the afternoon of July 3, 1863. They only have to ignore the paved roads, automobiles, and 1,328 monuments that cover the landscape.[55]

Taken together, these battlefield reconstructions suggest that while the war continues to shape terrain, its influence is limited by both nature and people. Antietam and Gettysburg survive in something resembling original form because they were and are pastoral landscapes, sites easily accessible to modern people who seek a closer connection with the war's seminal battles. That may not be a bad thing. Both Antietam and Gettysburg stand as idyllic examples of northern agriculture, with its emphasis on neatness, productivity and free labor. Such landscapes suggest, at least in part, the system of land use that the Union sought to preserve and extend. Both sites also allow visitors to understand the ways the material world—the geologic properties of limestone, the importance of the Gettysburg basin and the higher ground nearby—influenced tactics and command decisions. Other properties of those landscapes still remain elusive. At Antietam, only 1,937 of the battlefield's original 3,632 acres are open to visitors. The rest remains in private hands, with local landowners providing easements and restricting development on adjoining parcels. The popularity of Gettysburg as a tourist destination has turned the battlefield into something of a pastoral island, preserved amid the congested paved roads, hotels, fast food restaurants, and other amenities visible on its outskirts. At Fredericksburg and the Wilderness, those who would preserve the original battlefields have been far less successful in staving off both the whims of nature and the more systematic encroachment of suburban sprawl.

Even if the original physical landscapes could somehow be recovered, other environmental elements have been lost to history. The Park Service can never duplicate the exact weather conditions soldiers encountered. Visitors can never know the nervous anticipation of battle, the pain of a bullet wound, or the psychological stress of combat. Even the most dedicated Civil War reenactors, including those who fast in preparation for an event, cannot begin to experience the near-starvation and constant search for provisions typical of armies on the move. Nor would modern visitors relish the chronic diarrhea and other gastric ailments common on every battlefield. The unmistakable stench of decomposing bodies, the maggots and blowflies, the bloated carcasses of horses and mules are also nowhere to be found. Without these elements of nineteenth-century nature, even the most meticulously restored landscape can provide only a sanitized version of the terrain on which the struggle took place. Restored battlefields not only illustrate the Civil War's influence on today's landscape but also the ways modern Americans prefer to remember both the conflict and the natural world in which it took place.

EPILOGUE

AN ENVIRONMENTAL LEGACY

On April 9, 1865, Robert E. Lee met Ulysses S. Grant to discuss terms of surrender. The generals convened in the parlor of a house owned by Wilmer McLean, near a tiny community known as Appomattox Courthouse, Virginia. Those present at the event took pains to record nearly every detail of the meeting between the generals: the cordial initial conversation, Lee's courtly demeanor, Grant's benevolence, and the simplicity of the ceremony that ended four years of human carnage and sacrifice. In the months and years that followed, alleged eyewitness accounts of the meeting proliferated. Numerous artists tried to capture the extraordinary moment on canvas. Often exaggerated and embellished, such efforts helped sear the image of surrender into the American consciousness. Many of those in attendance scrambled for souvenirs. In short order, furniture, utensils, pencils, pens, scraps of paper, and other items in McLean's parlor disappeared.[1]

We know less about what happened outside the house on those days. Scattered observations suggest typical early spring weather, with fog in the morning hours of April 9 and rain later that night and the following day. In the various paintings and illustrations, trees appear bare or in early spring leaf. Searching for their own mementos, soldiers and bystanders cut limbs and bark from apple trees under which Lee allegedly stood or rested while in the area. Aside from those mangled trees — one of which even relinquished its roots to the souvenir hunters — events in nature seemed far less noteworthy than those unfolding in Wilmer McLean's parlor. So it has been for the Civil War as a whole. In the rush to explain the conflict's myriad effects on people and politics, the natural world has, until recently, largely escaped notice. But by the time Grant and Lee met at Appomattox, the Civil War had become an important environmental event, leaving a legacy that affected relationships between Americans and nature for decades to come.[2]

The long-term health of soldiers and civilians owed much to the disruption caused by the war. As soldiers flocked to camps in 1861, communicable diseases spread rampantly. But the war kept people in motion long after the initial musters and encampments, leaving unexposed populations vulnerable to disease. Following the Emancipation Proclamation in 1863, a second round of measles broke out as the Union army began to train African American troops, many of whom lacked immunity to *Rubeola*. As historian Jim Downs has shown, the half million slaves who escaped from bondage during the war and the 3.5 million African Americans freed by emancipation fared far worse than the soldiers. Smallpox, a disease that became more prevalent in the South during the first years of war, proved especially devastating. An epidemic that probably began among former slaves who fled to Union camps in Washington, D.C., in 1862 spread into the Deep South and the western territories by 1868, where it also wreaked havoc among Native Americans. The Freedmen's Bureau, a government agency established in 1865 to assist emancipated slaves, proved ill-equipped to deal with the epidemics and made matters worse by moving freed African Americans to regions in need of labor. As Downs argues, "The most significant factor that led to the outbreak of disease was the massive dislocation that the war and emancipation caused." Because of prevailing beliefs that former slaves were more susceptible to disease and less sanitary in their habits than white people, the federal government often failed to provide freedpeople with even the rudimentary medical care available at the time.[3]

White soldiers who survived various diseases during the war were less vulnerable to new infection, but they faced other problems. Before the war ended, Union medical authorities officially recorded nearly 80,000 cases of syphilis and more than 117,000 cases of gonorrhea. Thousands more went unreported. The devastating effects of those infections endured long after the shooting stopped, as venereal disease killed nearly one-third of all men who died in Union and Confederate veterans' homes. Likewise, historian Thomas Lowry writes, "No one knows how many Union and Confederate wives and widows went to their graves, rotted and ravaged by the pox that their men brought home, or how many veterans' children were blinded by gonorrhea or stunted by syphilis."[4]

During the long conflict, surgeons learned valuable lessons that radically altered and professionalized the nature of medical care in the late nineteenth century. The war presented them with opportunities to experiment with treatments of diseases and wounds, as well as institutionalize improved procedures and systems of care. As Shauna Devine argues, the field hospitals granted surgeons "the initiatives to develop investigative medicine through the study of gangrene" and other diseases, which "created a model for the integration of laboratory results and clinical observation." These practices that physicians developed while dealing with millions of injured and sick soldiers laid the foundation for the practice of modern specialized medicine.[5]

Those same techniques also left a long legacy of chemical dependency and addiction. In their treatment of wounds and sickness, Union surgeons distributed nearly 10 million opium pills to northern troops. So many soldiers took large doses of opium and its derivatives — as anesthetics or pain relievers — that many remained addicted to the drugs for decades. As Jonathan S. Jones argues, "Addiction ruined veterans' bodies, leaving them emaciated, impotent, and riddled with self-inflicted scars." It also made many of them ineligible for veterans' pensions. Opium addiction spread to the civilian population, too. One doctor wrote in 1868, "Maimed and shattered survivors from a hundred battle-fields, diseased and disabled soldiers released from hostile prisons, anguished and hopeless wives and mothers, made so by the slaughter of those who were dearest to them, have found, many of them, temporary relief from their sufferings in opium." By 1870, the northeastern portion of the United States consumed 500,000 pounds of opium, compared to only 105,000 pounds a decade earlier. By 1900, the United States had roughly 200,000 opium addicts.[6]

Soldiers lucky enough to avoid venereal disease and addiction discovered that a return to their farms did not guarantee a long or healthy postwar life. In 1892, John Shaw Billings, a former army medical inspector asserted that "the exertions, privations, and anxieties of military service . . . must necessarily have lowered the vitality and diminished the power of resistance to subsequent exposure and causes of disease." He conducted an extensive study of the vital statistics from Massachusetts and determined that, when compared to civilian males, veterans were twice as likely to develop tuberculosis or rheumatism, seven times more likely to suffer from heart disease, and fifty-five times more likely to experience diarrheal ailments. Civil War scholars have long known that for every soldier killed in battle, two more died of disease. We now know that disease, addiction, and chronic ailments were not simply wartime phenomena but rather part of a far more complex story of people and microbes that began in 1861 and went on long after hostilities ended.[7]

Though the war did not change North America's climate or atmospheric conditions, the conflict gave rise to a branch of the army that would eventually make the nation's weather much easier to monitor and predict. When the Civil War ended, Congress looked for ways to save on military appropriations. One proposal involved eliminating the Signal Corps, originally the brainchild of an assistant army surgeon named Albert J. Myer, who served with McClellan during the rain-soaked Peninsula Campaign. In 1863, Myer convinced the army to adopt his recently invented wigwag system of field communication. Using a single flag or torch and a preset system based on the telegraph code, trained soldiers could pass orders and other vital information from one site to another, a valuable communication tool during pitched battles like those of the Civil War. In peacetime, though, the Signal Corps seemed like a needless expense, and some politicians favored disbanding it.[8]

At the same time, Increase A. Lapham, a Wisconsin meteorologist and longtime observer of Great Lakes weather, began to lobby Congress for some national system to monitor the storms that interfered with shipping in the region. Before the Civil War, that responsibility had rested with the Smithsonian Institution, where scientists received and recorded observations from nearly 500 sites around the nation. Many of those outposts had been destroyed or abandoned during the conflict, and a fire at the Smithsonian in 1865 severely damaged its weather instruments. Championed by Wisconsin congressman Halbert E. Paine, Lapham's proposal eventually took the form of legislation "to provide for taking meteorological observations at the military stations in the interior of the continent and at other points" so that "the approach and force of storms" on the lakes and elsewhere could be communicated via "magnetic telegraph and marine signals." The measure created a new "Division of Telegrams and Reports for the Benefit of Commerce." It passed Congress as a joint resolution.[9]

As one of his first acts as president, Ulysses S. Grant, who had seen firsthand how weather could affect military operations, signed the resolution into law on February 9, 1870. Albert Myer immediately lobbied Paine to entrust the new duties to the Signal Corps. With the approval of the secretary of war, the corps became responsible for twenty-five weather stations across the nation that reported by telegraph to Washington, D.C. In 1890, Congress renamed the organization the Weather Bureau and moved it to the Department of Agriculture; in 1970, it became the National Weather Service.[10]

While the war left a lasting impact on health, medical practices, and weather reporting, it also shaped the relationship of animals and humans in the ensuing decades. In what might be the most fabled exchange between the two generals at Appomattox, Lee asked Grant if the Confederates might be allowed to keep their horses. The men would need them, Lee explained, for the spring plowing and planting. Though the provision was not part of the formal surrender terms, Grant agreed that any southern soldier claiming a horse or mule as his own could keep the animal. Historians often cite that concession as an example of Grant's benevolence, but the gesture had far-reaching environmental consequences for South and North alike.[11]

Some 1.2 million horses and mules died in service during the war—roughly 16 percent of the total number of equines in the country. The South bore the brunt of the loss. Indeed, by the time of the surrender, Grant had more horses in the field than Lee had men. By 1870, after five years of peace, Alabama, Arkansas, Georgia, North Carolina, South Carolina, and Virginia had only two-thirds as many horses and mules as in 1860. Only Texas, shielded from the worst of the fighting by sheer distance, saw its equine population increase. In contrast, some northern states, including Iowa, Minnesota, and Wisconsin, had more horses in 1870 than when the war began. Returning southern veterans did sorely need the horses they brought home from the war.[12]

The surrender made no provision for cattle, but those animals were now in short supply in certain regions. Foraging armies killed their share, but weather also affected the bovine population, especially in California. While the rainfall in the winter of 1862 had exceeded fifty inches along coastal regions stretching from San Francisco to Los Angeles, less than four inches fell in the winter of 1863, and even less fell the following winter. Water holes dried up, and thousands of cattle starved to death or suffocated from dust inhalation. Ranchers who had earned six cents a pound on beef before the drought now sold their skeletal animals for the value of their hides, getting as little as two dollars for an entire cow. In southern California, many ranchers drove their starving cattle over the cliffs at San Pedro south of Los Angeles as a "merciful solution" to their troubles. Nearly 30,000 head of cattle died on one ranch alone. The ranchers went bankrupt or sold their enormous landholdings for pennies on the dollar. In a dramatic reshaping of the region, much of greater Los Angeles developed on the land purchased from the failed cattle ranches. Altogether, while nearly 200,000 cattle drowned in the floods, an estimated 300,000 perished because of the drought, dropping that state's cattle numbers by 45 percent. It took thirty years for the herds to recover.[13]

Despite such heavy losses in the Far West, overall numbers of cattle remained steady in Union states during the war, suggesting that populations outside California continued to grow. In contrast, the Confederate states lost about 14 percent of their cattle between 1860 and 1870. Georgia and the Carolinas never again had as many cows roaming their woods and fields as in 1860. Even where cattle came back to prewar levels, rebuilding the herds took time. Only Florida, Mississippi, and Tennessee saw a return to prewar numbers by 1870. Arkansas and Texas did not recover until 1880; it took Alabama until 1890.[14]

Hogs fared the worst of all livestock. During the 1850s, the nation's swine population had increased more than 10 percent to 33.5 million animals; during the 1860s it fell by 25 percent, to a little more than 25 million. Once again, the losses weighed heaviest on the South. Reduction in hog imports from the Midwest, Union depredations, rampant hunger, and hog cholera led to a 70–80 percent decline in most southern states by the end of the war, with South Carolina suffering a staggering 85 percent loss. The South never recovered from the devastation. In 1860, the southern states had 15,562,800 hogs, representing almost 47 percent of the nation's total number. By 1900, the former Confederate states only had 15,055,483 hogs, a mere 24 percent of the nation's total. While Arkansas, Florida, Texas, and eventually Louisiana (by 1900) increased their total number of hogs, Alabama, Georgia, Mississippi, the Carolinas, Tennessee, and Virginia (including West Virginia) never produced as many hogs as they had in the year before the war began until the advent of factory hog farms in the latter half of the twentieth century. Hog numbers fell in some northern states as well—it took a lot of protein to keep the Union's 2 million soldiers fed—but Federal losses paled in comparison to those of the Confederacy. More important, a favorable corn-hog ratio meant that northern hog populations rebounded quickly after the war.[15]

During the last decades of the nineteenth century, the Midwest cemented its control over the pork trade. The South, which had just begun to import pork before the war, now became far more dependent on pigs produced elsewhere. As the gap between rich and poor widened and free-ranging cattle and hogs disappeared, southerners of all classes became ever more reliant on this imported pork. This "dependency in pork" became one of the war's enduring legacies, one that also adversely affected human health. Pork from pen-raised, corn-fed swine in the Midwest had a much higher fat content than meat from free-range hogs. Strange as it may seem, the Civil War might be partly responsible for the comparatively high rates of obesity, high blood pressure, stroke, and heart disease in the South, a trend that began in the nineteenth century and persists to the present, made worse by the massive factory farms that now dominate the southern hog industry.[16]

Just as with people, lethal microbes followed livestock off the battlefield back into civilian life. The exact movements of the various animals and diseases are difficult to track, but concern for livestock health began well before the war ended. With the creation of the Department of Agriculture (1862) and the American Veterinary Medical Association (1863), the federal government began to study the extent and effects of various animal epidemics. Shortly after the war, the reorganized army brought in two trained medical doctors to treat equine diseases, including glanders. In the mid-1860s, Department of Agriculture statistician Jacob Richard Dodge began publishing detailed reports of the many ailments affecting farm animals. Among the most prominent, he discovered, were cattle fever, glanders, and a variety of infections sometimes mistakenly lumped together as hog cholera. All had spread rapidly during and after the war, especially in newly settled regions. More ominously, by 1875, Dodge wondered if the various ailments affecting swine might soon make hogs an untenable source of human food.[17]

While those reports circulated, twenty-three of the nation's land grant colleges established courses of study that included animal health. By 1879, Iowa State had the first publicly supported college of veterinary medicine. According to medical historian G. Terry Sharrer, "the national government's involvement with the livestock industry began during the Civil War." Sharrer also suggests that development and professionalization of veterinary medicine and education can be traced directly to the conflict. As scholars delve further into the Civil War's impact on animals, the story of livestock and their assorted ailments may well prove as compelling as the migration of human diseases.[18]

Those soldiers who returned home after Appomattox with their horses were the lucky ones. Though the number of those who died in service remains difficult to calculate with precision, scholars estimate that as many as 750,000 soldiers perished in the war. The South lost at least 25 percent of men mobilized, while the North lost nearly 20 percent. Much research remains to be done, but available evidence suggests that the higher death toll in the Confederacy had important implications for the region's population dynamics. The dearth of young men made it more difficult for southern women of marriageable age to find an appropriate mate, especially during the 1870s. Some women postponed marriage. Others married older men or men of lesser social status. Southern males, who suddenly found themselves in higher demand, might have been more inclined to remarry if a spouse died. Others simply left their current wives, remarrying without a formal divorce. The lack of white males might also have contributed to the South's rigid postwar laws against miscegenation, as politicians worried that white women in search of husbands might turn more to freed black men. By the early 1880s, the "marriage squeeze" engendered by the war began to abate as a new male cohort took its place in the southern population. Roughly 92 percent of southern white women who reached marriageable age during the war eventually found a mate, but for a decade and a half, many of them lived with a heightened fear of spinsterhood.[19]

The absence of so many men also created a serious agricultural labor shortage. As with nearly every environmental consequence of the conflict, the South fared worse than the North. In the years immediately following the Civil War, total acres of cultivated land decreased throughout the region. In the eleven Confederate states, between 1850 and 1860, the amount of improved land had increased by 14.2 million acres, a gain of 33 percent. Between 1860 and 1870, the amount of improved land decreased by 9.85 million acres, a loss of 17 percent. Alabama had witnessed a 44 percent expansion in cultivated acres in the decade before the war, only to suffer a nearly 21 percent reduction during the war's decade. North Carolina's tilled acreage expanded nearly 20 percent before the war, only to lose nearly 20 percent in the following decade. However, the single greatest devastation in acreage belonged to the state that started the war. South Carolina saw its total tillable acreage go from a 12 percent increase by 1860 to a 34 percent decrease by 1870. The wartime losses were so severe that by 1880, these three states would still not be farming as many improved acres as they had before the war began.[20]

Predictably, production of food crops dropped sharply in the half decade after peace. Wheat production in the southern states increased by nearly 77 percent between 1850 and 1860 but fell by 22.5 percent by 1870. It took until 1880 before the southern states produced as much wheat as they had in 1860. Some states, like North and South Carolina, never again produced as much wheat as they had before the war. The southern staple, corn, fared even worse. The corn crop grew by 10 percent in the decade before the war in the Confederate states but dropped 34 percent by 1870. Every southern state (except Texas) lost between 20 and 55 percent of its corn crop by 1870. However, the states of Alabama, Georgia, Louisiana, Mississippi, and the Carolinas still did not produce as much corn in 1890 (twenty-five years after the war ended) as they had in 1860. This reveals not only the intensity of the destruction but also that farmers devoted the majority of their improved acres to cotton production in the decades following the war.[21]

Given the labor shortage, the diminished and sickly livestock population, and the devastation of fields and farms, one might wonder how the

South avoided complete ecological collapse. Fortunately, southern states and the rest of the nation had at their disposal something that helped stave off disaster—a vast supply of unoccupied land to the west. With the federal government forcibly removing Native Americans, providing access to land via the Homestead Act of 1862, and subsidizing the Transcontinental Railroad, westward expansion became the driving force in American economic growth during the last thirty years of the nineteenth century.

As southerners rebuilt old farms torn apart by war, previously uncultivated land—in the Mississippi Delta, Arkansas, Missouri, the upper Piedmont and lower Appalachians—provided the slowly recovering human population with fresh ground to till and ultimately helped save the region's agriculture. Such development required horse- and mule-power, and ambitious breeders in Tennessee, Missouri, and Kentucky—border states where recovery came quicker—rushed to fill the void. Texas also supplied horses, mules, and cattle to help restock southern farms. By 1880, Alabama, Georgia, and South Carolina still did not have as many equines as they had before the war, and the other southern states had barely increased their numbers. Texas mules helped plow the cotton fields that dominated the South (including Oklahoma). It took three decades, but by 1900 cotton exports from southern states were triple those of 1860, even though food crops had only just returned to prewar levels.[22]

Even so, expansion was no panacea for a region devastated by four years of relentless warfare. The Civil War destroyed approximately 4 million acres of southern forests, an area larger than the states of Connecticut and Rhode Island combined and a loss roughly equivalent to the deforestation in Vietnam during U.S. involvement there. That figure, though, pales in comparison to the fate of French forests in World War II, during which over 100 million acres of woodlands disappeared. Moreover, southern forests probably recovered quicker than those destroyed by modern weapons and chemical defoliants. Much of the timber harvested during the Civil War took the form of coppice cutting, in which hardwood stumps and smaller deciduous trees remained on the landscape. As soon as the armies moved on, saplings sprouted from the stumps, creating brushy forests like those at the Wilderness. Overall, pines might have become more common, as those species often invade abandoned fields in the Piedmont and coastal plain, a process that had been going on in one form or another from the time indigenous communities cleared the first fields in the region. As historian Megan Kate Nelson explains, the "landscapes of clearing and ghost forests" left by the war "were ultimately ephemeral." Because the worst of the deforestation followed armies deployed in the eastern theater, millions of acres of southern woodland remained untouched or only lightly affected by military demand. So much timber remained that northern and midwestern lumber companies moved into the South in the late nineteenth and early twentieth centuries—after the forests of the Great Lakes region had been exhausted.[23]

Recovering forests of shrubby hardwoods and pine were not the only signs of a changing postwar southern landscape. Before 1860, the South had built an agricultural system based on abundant land and enslaved labor that cleared new land and performed a series of land maintenance tasks that included ditching, weeding, manuring, and fencing fields. After Appomattox, white southerners of sufficient means began a relentless drive to produce cotton and other staples. This led to a dramatic increase in soil erosion from overworked and unrepaired fields, as emancipated slaves had no interest in performing back-breaking maintenance tasks without additional pay, and landowners stubbornly refused to pay them for it. As a result, landowners began fencing in their remaining cattle and hogs so that the animals might be more easily fattened for market or local consumption. Wealthier farmers also lobbied state legislatures to pass new laws that required such enclosures. Though the open range persisted for decades in parts of the South, such stock laws had the overall effect of denying poorer southerners, black and white, access to both land and animals. Lacking those necessities and without adequate federal assistance, former slaves and poorer white southerners became trapped in the postwar labor systems of tenant farming and sharecropping.[24]

The new system only intensified ecological problems that had long been part of southern agriculture. Soil exhaustion and erosion became hallmarks of the postwar landscape, especially in hillier regions. Not all the difficulties stemmed from planting cotton. As historian Drew Swanson has argued, in upland Virginia, the demand for bright leaf tobacco created during the war did not end when soldiers came home. As farmers began to exploit the new tobacco market, "soil erosion played havoc with the region's fragile hillsides, made worse by deforestation to fuel tobacco barns, and by the early twentieth century all of the topsoil was gone from broad swaths of the countryside surrounding Danville." By the 1930s, the newly created Soil Conservation Service declared the region "one of the nation's most serious environmental problems." Expansion to new ground might have allowed the South to avoid ecological collapse. But as historian Jack Temple Kirby explains, "The postwar landscape . . . relentlessly reorganized, deforested, and cotton-spread" eventually led to "a poorer South after all."[25]

While the South reorganized, the northern system of agriculture, with its emphasis on family farms and free labor, rolled onto the Great Plains. There and in the more arid Far West, free market agriculture would eventually create its own set of serious environmental problems. During the last years of the war, however, some Americans also looked to the West as a region that might help heal the nation. According to historian Aaron Sachs, with the return of peace, a number of American artists and intellectuals began to note "the grim parallel" between trees and men mangled in battle. Both had lost limbs or had once-healthy body parts reduced to stumps during four years of war. Almost anywhere one ventured in eastern America, healthy forests and healthy men had disappeared, while "the stumps multiplied unimaginably."[26]

It may be more than mere coincidence, then, that on May 17, 1864 — ten days after the Battle of the Wilderness and the beginning of the bloody Overland Campaign — Senator John Conness introduced a bill to protect Yosemite Valley and a stand of giant sequoias known as the Mariposa Grove by ceding them to the state of California. The huge trees, Conness explained, "have no parallel perhaps in the world" but needed protection as they were "subject now to damage and injury." Furthermore, the bill made clear that the scenery, especially the trees, had to be controlled by California "inalienable for all time." Instead of being reduced to stumps, like so many trees in the East, the sequoias had "to be used and preserved for the benefit of mankind."[27]

It probably did not hurt that the Unitarian minister Thomas Starr King, who had written extensively of Yosemite's beauty in 1860, had also been a staunch Unionist and had taken a leading role in staving off secession in California. But many who backed the bill saw other benefits in legislation to protect the valley and the trees. Frederick Law Olmsted, America's foremost landscape architect and head of a nine-man commission appointed to oversee Yosemite, believed contemplation of such grand scenery could greatly aid one's emotional health. "If this contemplation occurs in connection with relief from ordinary cares," Olmsted wrote, it could stave off "mental and nervous excitability, moroseness, melancholy, or irascibility" and increase "the subsequent capacity for happiness." Given the wanton destruction and wholesale mutilation of people and forests back east, one could scarcely have written a better prescription for what ailed the nation in 1865. Though his fellow commissioners eventually scrapped most of Olmsted's specific recommendations for Yosemite, the valley and the trees had been protected and, by almost any measure, the United States had taken an important step toward

Mirror Lake in Yosemite Park, 1865. With landscapes in the East torn asunder by war, Congress moved to protect the West's scenic wonders. Courtesy Library of Congress.

recognizing the healing properties of scenic landscapes, especially those in the American West that had remained unscarred by war.[28]

In the years that followed, John Muir—who had fled the United States to avoid service in the Union army and found both safety and solace in the Canadian wilderness—became Yosemite's chief publicist. Like Olmsted, Muir believed that nature afforded needed respite from the cares of everyday life, but he also urged Americans to turn to the natural world to find God and admire His handiwork. Those ideas had been percolating in American culture for several decades, spurred on by philosophers like Ralph Waldo Emerson and Henry David Thoreau. Artists and photographers had celebrated the western landscape as something uniquely American and a source of national pride. The movement toward protection of scenic landscapes was an enterprise fraught with irony, contradictions, and mixed motives. But the notions that one could be healed and find God in nature, ideas that came to the fore in a nation whose land and people had been torn asunder by war, proved a powerful motivation for setting aside other western landscapes. On June 30, 1864, President Lincoln signed the bill establishing Yosemite and preserved a stretch of land for improving the mental and physical health of the nation.[29]

The same year that John Conness rose in Congress to tout Yosemite's virtues, George Perkins Marsh published *Man and Nature; or, Physical Geography as Modified by Human Action*. Often regarded as the first environmental history, Marsh's work decried humanity's adverse effects on the natural world of the East, especially his native Vermont, and urged Americans to do better. The West provided that opportunity. Some of the tallest trees in the Mariposa Grove were later named for Civil War generals, including Robert E. Lee. In 1872, President Ulysses S. Grant, who knew better than most the effects of war on people and land, signed into law the bill that created Yellowstone National Park "for the benefit and enjoyment of the people." Echoing Olmsted, Stephen Mather, the first superintendent of the National Park Service, noted in 1920 that the national parks "contain the highest potentialities of national pride, national contentment, and national health." In 1933, the National Park Service took over administration of the most prominent Civil War battlefields. In what might be deemed its most important environmental legacy, the Civil War not only changed the American landscape but also forever altered how Americans thought about and began to take care of the natural world.[30]

ACKNOWLEDGMENTS

Before we became coauthors, we were baseball fans. In 2011, during a long ride home after taking in several games in Washington, D.C., we had our first conversation about this book. At the time we were each involved in separate projects related to the Seven Days' Battles of 1862 and our drive-time talk quickly turned to the role of weather in that critical campaign. Over the next year, usually during forays to nearby minor league parks, we talked more broadly about what an environmental history of the Civil War might involve. We were not, however, in any hurry to write such a book. In September 2012, while having lunch at the Boone Saloon, one of us (we agree it was probably Judkin, but we can't be sure) began to scribble chapter titles and major themes on a cocktail napkin. It took another half decade for the manuscript to begin to take shape. Along the way we had help from a variety of institutions and individuals without whom our casual conversations and intermittent scribblings would never have made it into print.

In the Department of History at Appalachian State University, we had the good fortune to work with two supportive department chairs, Lucinda McCray and James Goff, who did what they could to arrange our schedules and lighten our teaching loads so that we had time for scholarship. Other colleagues, too, offered support and friendly critiques of our work at faculty seminars. Even so, we might still be pounding our keyboards had it not been for the enormous generosity of the American Council of Learned Societies. In 2013, the council awarded us a prestigious Collaborative Research Fellowship that allowed us to work full time on the project during the 2014–15 academic year. In addition, Appalachian State awarded us its 100 Scholars Research Grant that helped facilitate travel and research. Zachary Hottel provided us with some of our earliest research into the effects of mud.

Chris Robey volunteered his services as an undergraduate research assistant, working for free until we found a way to get him paid.

A number of critics and commentators helped sharpen our arguments as we floated trial balloons of our unfinished chapters at various conferences. Tim debuted some of our ideas at "The Blue, the Gray, and the Green," the first academic conference devoted solely to Civil War environmental history, in Athens, Georgia, in 2011. We particularly appreciate the efforts of Joan Cashin, Mart Stewart, and Lisa M. Brady, who graciously allowed us to join their panel on the Civil War at the 2014 meeting of the American Society for Environmental History in San Francisco. We also thank Kathryn Shively Meier for inviting us to be commenter and chair of an environmental panel at the 2014 Society of Civil War Historians meeting in Baltimore (not least because the venue was within walking distance of Oriole Park at Camden Yards, where we took our evening meals). Various other conferences provided us with the opportunity to hear innovative papers from a number of talented younger scholars working on the war and the environment, including Michael Burns, Jonathan Jones, Erin Stewart Mauldin, Lindsay Rae Privette, and David Schieffler. We also leaned heavily on the published work of myriad other historians who have considered the Civil War in the context of the natural world. Our debt to them is acknowledged in the notes. The late Jack Temple Kirby merits special mention, as his essay for TeacherServe at the National Humanities Center has prompted many would-be environmental historians—present company included—to gravitate toward the Civil War.

Mark Simpson-Vos of the University of North Carolina Press has served as our editor, always urging us to write for a general audience and constantly reminding us that readers, especially those interested in the Civil War, appreciate a linear narrative. He was also gracious and understanding when we needed a deadline extension. Alex Martin expertly copyedited our manuscript. We thank Kenneth Noe for reading an early draft of our second chapter and making valuable suggestions. Lisa M. Brady, Erin Stewart Mauldin, and Megan Kate Nelson read the entire manuscript and provided comments and criticisms that greatly improved the final product. Peter Carmichael also offered a strong endorsement after reading our draft. Some of our work on the Battle of Antietam appeared in an essay by Lisa M. Brady and Timothy Silver, "Nature as Material Culture: Antietam National Battlefield," in *War Matters: Material Culture in the Civil War Era*, edited by Joan Cashin (Chapel Hill: University of North Carolina Press, 2018), 53–74. Likewise, much of our work on the Peninsula Campaign first appeared in Judkin Browning and

Timothy Silver, "Nature and Human Nature: Environmental Influences on the Union's Failed Peninsula Campaign, 1862," *Journal of the Civil War Era* 8 (September 2018): 388–415. We are grateful for permission to reuse material from both essays here.

One of the potential perils of environmental history is that it requires those trained in the humanities to venture into the sciences where, as scholars, we are proverbial strangers in a strange land. Fortunately, we had patient and affable guides as we wandered through unfamiliar disciplines. From Appalachian State's Department of Geology, Andrew B. Heckert directed us to important sources and Cynthia M. Liutkus-Pierce helped us understand the nature of soil and water tables in the Peninsula Campaign. Brian Ledford, the area forester for Smyth County for the Virginia Department of Forestry, provided information on different types of trees and their growth rates. Bert Way, of Kennesaw State University, directed us to some important sources on forestry. Dr. Robert Griffith, MD, served as an informal medical consultant. Penny Blevins at the Museum of Middle Appalachia in Saltville, Virginia, was exceptionally helpful, as were intern Ken McCracken and park ranger Frank Lloyd at the Wilderness Shelter and Spotsylvania battlefield shelters when we visited in 2018.

Coauthored books are not common among historians, and more than once over the past few years, we came to understand why. Our friendship and affinity for baseball notwithstanding, we have argued, cajoled, and put up with each other's scholarly quirks as we wrestled with every chapter. Though we individually drafted parts of the manuscript, we planned and revised it together, laboring over every line of prose. This book is truly a collaborative effort, one that neither of us could have managed alone. That we emerged from such an endeavor with our friendship intact is testament to the respect that we developed for each other during that long and tedious process. Our families also lived with this project and offered nothing but encouragement. Judkin thanks his wife, Greta, and daughter, Bethany, for their support and willingness to pretend they were not revolted by his dinner-table discussions of such topics as diarrhea, horse manure production, and the mechanics of starvation or dying of thirst. Tim thanks his wife, Cathia, for her support and careful reading of the manuscript and his daughter, Julianna, who was never too busy with her own college life to ask about Dad's work. They all deserve recognition as vital partners in our efforts.

NOTES

INTRODUCTION

1. Rable, *Fredericksburg! Fredericksburg*, 389–410.
2. Winters et al., *Battling the Elements*, 36–39.
3. Trobriand, *Four Years with the Army of the Potomac*, 406–8; Armistead, *Horses and Mules in the Civil War*, 56; Dodge, *On Campaign with the Army of the Potomac*, 154.
4. Dodge, *On Campaign with the Army of the Potomac*, 154; Plumb, *Your Brother in Arms*, 67.
5. For more on the nature of Union military leadership, see Williams, *Lincoln and His Generals*; Weigley, *A Great Civil War*; and Murray and Hsieh, *A Savage War*.
6. Winters et al., *Battling the Elements*, 33–39, 44. For more on analysis of Burnside, see Williams, *Lincoln and His Generals*; Murray and Hsieh, *A Savage War*; Marvel, *Burnside*.
7. Kirby, "The American Civil War"; Brady, *War upon the Land*. Other scholars who have taken up the study of the war and nature are Steinberg, *Down to Earth*; Humphreys, *Marrow of Tragedy*; Bell, *Mosquito Soldiers*; Nelson, *Ruin Nation*. On the role of the environment in strategy and campaigns, see Brady, *War upon the Land*; Fiege, *The Republic of Nature*, 199–227; Drake, *The Blue, the Gray, and the Green*; Noe, "Heat of Battle"; and Meier, *Nature's Civil War*. For more on the environmental effects of the war on agriculture, see Mauldin, *Unredeemed Land*.
8. Sutter, "Epilogue: Waving the Muddy Shirt," 228–29; Crosby, *Ecological Imperialism*, 270.
9. Stroud, "Does Nature Always Matter?," 80.
10. Duncan, "The Campaign of Fredericksburg"; Lonn, *Desertion during the Civil War*, 132; Barney, *The Oxford Encyclopedia of the Civil War*, 98. Numerous soldiers filed pension claims after the war declaring that rheumatism from the Mud March wrecked their health. For a sample, see James N. Bartlett, Nathaniel Halstead, and Jerome Stockham pension files, Twenty-Fourth Michigan Infantry, Record Group 15, Federal Pension Application Files, Records of the Veterans Administration, National Archives, Washington, DC.
11. Sutter, "Epilogue: Waving the Muddy Shirt," 228; Berry, "Forum."
12. Robillard, *The Poems of Herman Melville*, 9.

CHAPTER 1

1. "The Feeling in Edgecombe"; "Southern Rights Meeting"; Manarin and Jordan, *North Carolina Troops*, 5:494–605. The first enlisting company became Company K, while the latter-enlisting company became Company I of the Fifth North Carolina Volunteers, later to be redesignated as the Fifteenth North Carolina Infantry. A company typically consisted of roughly 100 men, most of whom hailed from the same localities (counties or districts within cities). Typically, ten companies formed a regiment.

2. J. A. McM. to Editor, August 24, 1861, in *Raleigh (NC) Semi-weekly Standard*, September 4, 1861, 2; Centers for Disease Control, "Measles (Rubeola)." Our list of symptoms among the recruits is based on the typical course of measles infection.

3. Humphreys, *Marrow of Tragedy*, 92; Onion, "Late Nineteenth-Century Maps"; Centers for Disease Control and Prevention, "Measles (Rubeola)."

4. Humphreys, *Marrow of Tragedy*, 92; Hamidullah, "The Impact of Disease on the Civil War."

5. Hattaway, *Reflections of a Civil War Historian*, 180; Wiley, *The Life of Johnny Reb*, 251; Morgan, *Personal Reminiscences of the War*, 86; Sharrer, *A Kind of Fate*, 6, 387.

6. Report of Surgeon Charles S. Tripler, February 7, 1863, U.S. War Department, *The War of the Rebellion*, ser. 1, 5:82 (hereafter *War of the Rebellion*); U. S. Grant to General S. Williams, November 14, 1861, *War of the Rebellion*, ser. 1, 3:570; Henry Halleck to George McClellan, December 6, 1861, *War of the Rebellion*, ser. 1, 8:409.

7. Humphreys, *Marrow of Tragedy*, 92.

8. Sharrer, *A Kind of Fate*, 387; Woodward, *Outlines of the Chief Camp Diseases*, 268.

9. Barnes, *The Medical and Surgical History of the War of the Rebellion*, 1:30; "Civil War Medical Terms"; Adams, *Living Hell*, 23.

10. Sherwood, *Human Physiology*, 651.

11. U.S. Department of Agriculture, "Dietary Reference Intakes."

12. Montague, "Subsistence of the Army of the Valley," 227.

13. Gillet, *The Army Medical Department*, 159. Multiple different cuts of beef average approximately 50 calories per ounce (or 1,000 calories per ration allowance), but much fattier bacon averages approximately 150 calories per ounce (or 1,800 calories per ration allowance). Similarly, flour—whether white or buckwheat—averages just under 100 calories per ounce (approximately 1,500 calories per ration allowance), while cornmeal, regardless of variety, averages 90 calories per ounce (approximately 1,800 calories per ration allowance). All calories estimated using www.CalorieKing.com.

14. Cunningham, *Field Medical Services at the Battle of Manassas*, 6.

15. Piston and Hatcher, *Wilson's Creek*, 85, 120, 131, 148, 168.

16. Cunningham, *Field Medical Services at the Battle of Manassas*, 10; Lipman, "Bull Run Battle 1861."

17. Piston and Hatcher, *Wilson's Creek*, 100–105, 144; Knox, *Camp-Fire and Cotton Field*, 69.

18. Nelson, "'The Difficulties and Seductions of the Desert,'" 37–39; Josephy, *The Civil War in the American West*, 52.

19. Nelson, "The Difficulties and Seductions of the Desert," 41–44.

20. Nelson, "The Difficulties and Seductions of the Desert," 43–44.

21. Manarin and Jordan, *North Carolina Troops*, 5:494–616, 5:589; George to wife, August 1, 4, and 31, 1861, George Loyall Gordon Papers, Southern Historical Collection,

University of North Carolina at Chapel Hill; J. Bankhead Magruder to Col. George Deas, August 9, 1861, and Magruder to General S. Cooper, September 7, 1861, *War of the Rebellion*, ser. 1, 4:572, 644.

22. Freemon, *Gangrene and Glory*, 205–6. Modern doctors prefer the term "enteric" over "typhoid" fever because *Salmonella paratyphi* is also a source of the disease. Here we use "typhoid fever" as a generic descriptor to cover both diseases. Mayo Clinic Staff, "Diseases and Condition"; Alvarez-Ordonez et al., "Salmonella spp. Survival Strategies," 3268–81.

23. Earle, "Environment, Disease, and Mortality in Early Virginia," 370–72; Cohen, "Did Jamestown Settlers Drink Themselves to Death?"

24. J. Bankhead Magruder to Col. George Deas, August 9, 1861, *War of the Rebellion*, ser. 1, 4:572; C. H. Richardson to Dr. N. S. Crowell, August 8, 1861, *War of the Rebellion*, ser. 1, 4:632–33. For more on the prevalence of illness on the Virginia Peninsula, and how soldiers dealt with it, see Meier, *Nature's Civil War*.

25. Entry dated October 19, 1861, in Stone, *Brokenburn*, 62; Thomas E. Bramlette to George Thomas, December 23, 1861, *War of the Rebellion*, ser. 1, 7:513.

26. Sharrer, *A Kind of Fate*, 7; Aryal, "Differences between Diarrhea and Dysentery"; Humphreys, *Marrow of Tragedy*, 98; Schmidgall, *Intimate with Walt*, 187.

27. Centers for Disease Control and Prevention, "Question and Answers"; Mayo Clinic Staff, "E. Coli"; Mayo Clinic Staff, "Salmonella Infection."

28. Report of Surgeon Charles S. Tripler, February 7, 1863, *War of the Rebellion*, ser. 1, 5:83.

29. Wm. S. Holmes to Dear Friends at Home, August 4, 1861, in New York State Military Museum and Veterans Research Center, "35th Regiment New York Volunteer Infantry"; Adams, *Living Hell*, 20; Meier, *Nature's Civil War*.

30. J. F. Hammond, "Extracts from a Narrative of His Service in the Peninsular Campaign," in Barnes, *The Medical and Surgical History of the War of the Rebellion*, 2:65; A. Watson to Surgeon General, August 31, 1863, and A. Hartsuff to Surgeon General, September 29, 1863, in Barnes, *The Medical and Surgical History of the War of the Rebellion*, 3:50; Joseph Janvier Woodward, "Diarrhea and Dysentery" (1879), in Barnes, *The Medical and Surgical History of the War of the Rebellion*, 4:626.

31. Taylor, "Some Experiences of a Confederate Assistant Surgeon," 105.

32. "Calomel"; Morris, *The Better Angel*, 93.

33. Mayo Clinic Staff, "Syphilis"; Mayo Clinic Staff, "Gonorrhea"; Hart, "Sexual Behavior in a War Environment," 218–20.

34. Baxter, *Statistics, Medical and Anthropological*, 142; Onion, "Late Nineteenth-Century Maps"; U.S. Bureau of the Census, "Population of the 100 Largest Urban Places"; Adams, *Living Hell*, 21–22; Frank, *The World of the Civil War*, 1:174.

35. Clinton, "Public Women," 62; Wiley, *The Life of Johnny Reb*, 54–56.

36. Adams, *Living Hell*, 21; Schroeder-Lein, *Encyclopedia of Civil War Medicine*, 323.

37. Ash, *When the Yankees Came*, 85–86.

38. Weiss, "Dirty Water, People on the Move," 189–90; Bray, *Armies of Pestilence*, 155.

39. Bray, *Armies of Pestilence*, 155–70; Rosenberg, *The Cholera Years*, 13–174; Pyle, "The Diffusion of Cholera in the United States," 69; Wilford, "How Epidemics Helped Shape the Modern Metropolis." Pyle says that 15,219 New Yorkers died of cholera or symptomatic related maladies, while Wilford says that 5,071 people died.

40. Ponting, *The Crimean War*, 334; Smallman-Raynor and Clift, "The Geographical Spread of Cholera," 32–69; Rosenberg, *The Cholera Years*, 175–234.

41. World Health Organization, "Typhus Fever"; Occhiuti and McLelland, "Typhus Overview." In 1489, typhus killed nearly 17,000 of the 25,000-man Spanish army that tried to drive the Moors from their fortress at Granada on the Iberian Peninsula. At the siege of Prague in 1741, 30,000 Prussian troops succumbed to the infection, and the German army had to rethink its military and medical strategies in an effort to slow down the disease. When Napoleon invaded Russia in 1812, typhus (along with dysentery) killed half of his 600,000-man Grande Armée and led directly to his defeat. As his soldiers retreated through Europe in the winter of 1812–13, they spread the disease through the towns they passed, killing nearly 300,000 people in the German territories. Conlon, "The Historical Impact of Epidemic Typhus," 8–12; Dodman, "1814 and the Melancholy of War," 36.

42. Humphreys, "A Stranger to Our Camps," 270–71; Poe, *True Tales of the South at War*, 2–3; Miller, "Historical Natural History"; Adams, *Living Hell*, 47.

43. Humphreys, "A Stranger to Our Camps," 281–82.

44. Humphreys, "A Stranger to Our Camps," 285–88. Frank Freemon argues that Civil War soldiers bathed too often for lice to be a carrier of the disease, but Humphreys disagrees with that interpretation. Freemon, *Gangrene and Glory*, 206.

45. Silver, *A New Face on the Countryside*, 155–59; Humphreys, *Malaria*, 37; Bell, *Mosquito Soldiers*, 11; Schroeder-Lein, *Encyclopedia of Civil War Medicine*, 192–93. For more on the ecology and effects of mosquitoes in wartime, see also McNeill, *Mosquito Empires*.

46. Silver, *A New Face on the Countryside*, 155; Bell, *Mosquito Soldiers*, 10–11; Centers for Disease Control and Prevention, "Human Factors and Malaria"; Humphreys, *Malaria*, 37.

47. Silver, *A New Face on the Countryside*, 159–61.

48. Humphreys, *Malaria*, 37.

49. Humphreys, "This Place of Death," 14–15, 20; General Orders No. 9, September 9, 1861, *War of the Rebellion*, ser. 1, 5:95.

50. T. W. Sherman to Gen. Meigs, December 10, 1861, *War of the Rebellion*, ser. 1, 4:202–3. For more on the new diseases that freed people brought into army camps, see Downs, *Sick from Freedom*; and Taylor, *Embattled Freedom*.

51. Hong, "The Burden of Early Exposure to Malaria," 1001, 1002, 1011; Humphreys, *Malaria*, 37; Humphreys, *Intensely Human*, 48; Bell, "'Gallinippers' & Glory," 379–405.

52. "Tropical Plant Database"; Report of Surgeon Charles S. Tripler, February 7, 1863, *War of the Rebellion*, ser. 1, 5:83; R. R. Clarke to Metcalf, June 5, 1862, Box 3, Folder 5, Civil War Collection, American Antiquarian Society; Bell, *Mosquito Soldiers*, 31, 115; Villacorta, "The Civil War Wounded in Photographs."

53. Bell, *Mosquito Soldiers*, 109, 110, 115.

54. Humphreys, "This Place of Death," 17; Bell, *Mosquito Soldiers*, 15; Silver, *A New Face on the Countryside*, 162.

55. Dobson, *Murderous Contagion*; Mackowiak and Sehdev, "The Origin of Quarantine," 1071–72; Humphreys, *Yellow Fever and the South*.

56. Grav, "When the Beast Saved the Day," 37–41.

57. Grav, "When the Beast Saved the Day," 38; Humphreys, *Marrow of Tragedy*, 95–96.

58. Bell, "'Gallinippers' & Glory," 379–405.

59. Bell, *Mosquito Soldiers*, 104–6, 108–9; Roberts, "Yellow Fever Fiend."

60. Bell, *Mosquito Soldiers*, 104–6, 108–9; Roberts, "Yellow Fever Fiend."

61. Fenn, *Pox Americana*, 15–20.

62. Fenn, *Pox Americana*, 31–33.

63. Report of Surgeon Charles S. Tripler, February 7, 1863, *War of the Rebellion*, ser. 1, 5:84–85.

64. Humphreys, *Marrow of Tragedy*, 284, 285–86; Zapata, "How Civil War Soldiers Gave Themselves Syphilis."

65. Fenn, *Pox Americana*, 88–91.

66. Henry Payton to Beauregard, May 25, 1862, *War of the Rebellion*, ser. 2, 3:556; Beauregard to Henry Halleck, May 20, 1862, *War of the Rebellion*, ser. 2, 5:150; H. W. Freedley to William Hoffman, January 3, 1863, *War of the Rebellion*, ser. 2, 5:150; William Hoffman to Lt. Col. F. A. Dick, February 16, 1863, *War of the Rebellion*, ser. 2, 5:277; Humphreys, *Marrow of Tragedy*, 285; "North Alton Confederate Cemetery."

67. Humphreys, *Intensely Human*; Downs, *Sick from Freedom*.

CHAPTER 2

1. Burt, "California's Superstorm"; Masich, *The Civil War in Arizona*, 9.

2. "The Great Flood in California"; Ingram, "California Megaflood"; entries dated January 31 and February 9, 1862, in Brewer, *Up and Down California in 1860–1864*, 243, 244; Porter et al., "Overview of the ARkStorm Scenario," 1–2.

3. "The Great Flood in California"; Porter et al., "Overview of the ARkStorm Scenario," 2, 71. In February 1862 in southern California, ranchers were charging the U.S. Army $25 per head of cattle. That price multiplied by 200,000 equals $5 million. G. R. Riggs to James H. Carleton, February 15, 1862, in U.S. War Department, *The War of the Rebellion*, ser. 1, vol. 50, pt. 1:869–70 (hereafter *War of the Rebellion*).

4. G. Wright to James H. Carleton, January 31, 1862, in *War of the Rebellion*, ser. 1, 4:90–91; Josephy, *The Civil War in the American West*, 269–72.

5. "Humid Continental Climate."

6. "Natural Climate Oscillations"; Weier, "El Niño's Extended Family"; "El Niño & La Niña."

7. Herweijer, Seager, and Cook, "North American Droughts," 160–61.

8. Guan et al., "Does the Madden-Julian Oscillation Influence Wintertime Atmospheric Rivers?," 325–42; NOAA, "What Is the Pineapple Express?"

9. Porter et al., "Overview of the ARkStorm Scenario," 1–2; NOAA, "Earth System Research Laboratory"; Burt, "California's Superstorm."

10. "What Is the Rain Shadow Effect?"

11. Josephy, *The Civil War in the American West*, 52; Frazier, *Blood and Treasure*, 75.

12. Josephy, *The Civil War in the American West*, 76–80; Frazier, *Blood and Treasure*, 117–230.

13. Josephy, *The Civil War in the American West*, 85–92.

14. Thompson, *Confederate General of the West*, 226–32.

15. Davidson, "Battle of Valverde," 31; Clough, "Battle of Valverde," 46, 53.

16. Thompson, *Confederate General of the West*, 236; Davidson, "From Socorro to Glorieta," 79; Davidson, "Dona Ana to San Antonio," 128.

17. Davidson, "Dona Ana to San Antonio," 127–29.

18. U.S. Geological Survey and Tennessee Valley Authority: Mapping Services Branch, *Shooks Gap Quadrangle*; U.S. Geological Survey and Tennessee Valley Authority, Mapping Services Branch, *Paducah East Quadrangle*; Coggins, *Tennessee Tragedies*, 259–62.

19. Albert Sidney Johnston to Judah P. Benjamin, November 8, 1861, in *War of the Rebellion*, ser. 1, 4:128; Grant, *Personal Memoirs*, 1:175; Gott, *Where the South Lost the War*, 61–62.

20. Cooling, *Forts Henry and Donelson*, 79–80, 88, 103; Kemmerly, "Environment and the Course of Battle," 1081–83; Lloyd Tilghman to S. Cooper, February 12, 1862, in *War of the Rebellion*, ser. 1, 7:139.

21. Cooling, *Forts Henry and Donelson*, 122, 147, 200; Woodworth, *Nothing but Victory*, 85–93.

22. Woodworth, *Nothing but Victory*, 115–18.

23. S. L. Phelps to A. H. Foote, February 10, 1862, in *War of the Rebellion*, ser. 1, 7:154; McDonough, *Shiloh*, 6–11.

24. McDonough, *Shiloh*, 19, 37–46; Grant, *Personal Memoirs*, 1:218, 198.

25. McDonough, *Shiloh*, 69–85, 152–55; Grant, *Personal Memoirs*, 1:206; Woodworth, *Nothing but Victory*, 141–206; Kemmerly, "Environment and the Course of Battle," 1086–1105.

26. Salling, *Louisianans in the Western Confederacy*, 35; letters dated May 5 and August 13, 1862, in Callaway, *The Civil War Letters of Joshua K. Callaway*, 10, 49.

27. Smith, *The Untold Story of Shiloh*, 75–76.

28. "La Niña Fact Sheet."

29. "Little Ice Age"; Fagan, *The Little Ice Age*.

30. Krick, *Civil War Weather in Virginia*, 2, 15, 116, 45, 30, 135.

31. "Chaos"; Silver, "The Weather Man Is Not a Moron."

32. Browning, *The Seven Days' Battles*, 13–21.

33. Sears, *To the Gates of Richmond*, 4–18; George B. McClellan to Edwin Stanton, February 3, 1862, in *War of the Rebellion*, ser. 1, 5:45.

34. Miller, "I Only Wait for the River," 47–48; George B. McClellan to Winfield Scott, April 11, 1862, and McClellan to Ambrose Burnside, May 21, 1862, in McClellan, *The Civil War Papers*, 236, 269.

35. Monette and Ware, "Early Forest Succession," 80–86; Dougherty, *The Peninsula Campaign*, 62–63; Sears, *To the Gates of Richmond*, 36.

36. Miller, "Weather Still Execrable," 193–94.

37. Sears, *To the Gates of Richmond*, 4–25; Sears, *George B. McClellan*, 168.

38. Browning, *The Seven Days' Battles*, 17–18; Dougherty, *The Peninsula Campaign*, 62; Shulten, "Mismapping the Peninsula." For more on the maps of the peninsula at the time, see Woolridge, *Mapping Virginia*, 291–303.

39. Sears, *To the Gates of Richmond*, 18–25; McLaws, *A Soldier's General*, 136–37.

40. Dougherty, *The Peninsula Campaign*, 62–63, 72–73, 78; Report of Major General George B. McClellan, August 4, 1863, in *War of the Rebellion*, ser. 1, vol. 11, pt. 1:12.

41. Sears, *To the Gates of Richmond*, 24–25, 35–39; Browning, *The Seven Days' Battles*, 18, 20–21.

42. Diary entry for April 6, 1862, in Sneden, *Eye of the Storm*, 40.

43. Miller, "Weather Still Execrable," 180–82; Burton, *Extraordinary Circumstances*, 6.

44. Jones, "The Influence of Horse Supply," 360–61; George B. McClellan to Edwin Stanton, May 5, 1862, in McClellan, *The Civil War Papers*, 255. Megan Kate Nelson notes that armies had to cut trees along roadways in order to widen the lanes or create detours, or to corduroy the roads. This removal of nearby trees only exacerbated future muddy

conditions as the area eroded quickly and could not control rain runoff. Nelson, *Ruin Nation*, 113.

45. Soper, *The "Glorious Old Third,"* 158; Sternhell, *Routes of War*, 54; Tally Simpson to Anna Tallulah Simpson, May 13, 1862, in Simpson, *"Far, Far from Home,"* 123.

46. For average Civil War soldier weight, see Brennan, "The Civil War Diet," 38. For metabolic equivalent (MET) expenditures, see "2011 Compendium of Physical Activity." We determined that "marching rapidly, military, no pack" was a reasonably similar expenditure to marching through a viscous surface with a pack. Other possibilities also suggest an 8 MET is appropriate. We thank Zachary Hottel (MA, public history, 2015 from Appalachian State University) for his assistance in determining the rate of energy expended. Much of this information derived from a graduate seminar paper that Hottel wrote in April 2014.

47. McClellan to Ellen, May 6, 1862, McClellan to Stanton, May 5, 1862, in McClellan, *The Civil War Papers*, 257, 255; entry dated May 6, 1862, in Berkeley, *Four Years in the Confederate Artillery*, 17.

48. Johnson, *Geology of the Yorktown*, 3–41, plate 1.

49. Johnson, *Geology of the Yorktown*, 3–41, plate 1; entry dated May 6, 1862, in Wainwright, *A Diary of Battle*, 58. We thank Cynthia Liutkus-Pierce of Appalachian State University's Department of Geology for helping us to understand the geologic history of the region and the nature of the different types of sedimentary layers.

50. McClellan to Winfield Scott, April 11, 1862, McClellan to Stanton, May 30, 1862, and McClellan to Ellen, May 5, 1862, in McClellan, *The Civil War Papers*, 236, 281, 255; Miller, "The Grand Campaign," 180.

51. Miller, "I Only Wait for the River," 47–55.

52. Browning, *The Seven Days' Battles*, 40; Report of General George McClellan, *War of the Rebellion*, ser. 1, vol. 11, pt. 1:41.

53. Berkeley, *Four Years in the Confederate Artillery*, 18; Dowdey, *The Seven Days*, 84–127.

54. Miller, "Weather Still Execrable," 185–86; McClellan to Mary Ellen, June 10, 1862, in McClellan, *The Civil War Papers*, 294; Robert E. Lee to Jefferson Davis, June 5, 1862, in Lee, *The Wartime Papers*, 184.

55. Browning, *The Seven Days' Battles*, 27–59; Sears, *To the Gates of Richmond*, 345.

56. Entry dated June 29, 1862, in Sneden, *Eye of the Storm*, 82; Report of Surgeon Jonathan Letterman, March 1, 1863, in *War of the Rebellion*, ser. 1, vol. 11, pt. 1:214. See also Stewart, "Walking, Running, and Marching."

57. Browning, *The Seven Days' Battles*, 86, 104, 152; Gerleman, "Unchronicled Heroes"; entry dated June 29, 1862, in Sneden, *Eye of the Storm*, 82.

58. Sharrer, *A Kind of Fate*, 8; Trobriand, *Four Years with the Army of the Potomac*, 254–55; Wilbur Fisk letter, January 15, 1863, in Fisk, *Hard Marching Every Day*, 44.

59. McClellan, *The Civil War Papers*, 282; Bell, *Mosquito Soldiers*, 74; Sharrer, *A Kind of Fate*, 8.

60. Report of Surgeon Charles S. Tripler, June 17, 1862, in *War of the Rebellion*, ser. 1, vol. 11, pt. 1:207–10.

61. Report of Surgeon Jonathan Letterman, March 1, 1863, *War of the Rebellion*, ser. 1, vol. 11, pt. 1:210–15; McClellan to Mary Ellen, July 11, 1862, in McClellan, *The Civil War Papers*, 351.

62. Barnes, *The Medical and Surgical History of the War of the Rebellion*, 1:174–79; Report

of Surgeon Jonathan Letterman, March 1, 1863, *War of the Rebellion*, ser. 1, vol. 2, pt. 1:210–12; Bell, "'Gallinippers' & Glory," 389.

63. Burton, *Extraordinary Circumstances*, 340–50; Browning, *The Seven Days' Battles*, 131–51; Sears, *To the Gates of Richmond*, 281, 337–38.

64. Sears, *To the Gates of Richmond*, 352–53.

65. McClellan to Mary Ellen, July 8 and August 8, 1862, in McClellan, *The Civil War Papers*, 346, 387.

66. Browning, *Shifting Loyalties*, 128; Noe, *Perryville*, 24, 81; Harris, *Piedmont Farmer*, 252–58.

67. Noe, *Perryville*, 24–25.

68. Noe, *Perryville*, 22–41.

69. Noe, *Perryville*, 63–73.

70. Noe, *Perryville*, 74–98, 106.

71. Foote, *The Civil War*, 1:729; Noe, *Perryville*, 136, 120; Obreiter, *The Seventy-Seventh Pennsylvania*, 128.

72. Noe, *Perryville*, xiv, 215, 313–39; Obreiter, *The Seventy-Seventh Pennsylvania*, 131.

73. Clausewitz, *On War*, 66. For an analysis of Clausewitz and friction, see Brady, "Nature as Friction."

74. Sutter, "The World with Us," 94–119; Drake, "New Fields of Battle," 7; Nash, "The Agency of Nature."

75. For the religious nature of American society, see Rable, *God's Almost Chosen People*.

CHAPTER 3

1. Sutherland, *Seasons of War*, 5.

2. U.S. Census Bureau, "1860 Census: Population of the United States"; Worster, "Transformations of the Earth," 1095–96; Stoll, *Larding the Lean Earth*, 15–16; Sutherland, *Seasons of War*, 5.

3. Grimsley, *The Hard Hand of War*, chaps. 2–5; letter dated June 7, 1862, in Voris, *A Citizen Soldier's Civil War*, 60.

4. Hennessy, *Return to Bull Run*, 1–38.

5. McPherson, *Battle Cry of Freedom*, 499–502; Grimsley, *The Hard Hand of War*, chaps. 2–5.

6. Sutherland, *Seasons of War*, 5, 123, 178; McSween, *Confederate Incognito*, 68.

7. Montague, "Subsistence of the Army of the Valley," 229–30. For a discussion of scholarly explanations of Jackson's poor performance, see Browning, *The Seven Days' Battles*, 37–38.

8. Hennessy, *Return to Bull Run*, 116–30; Robert E. Lee to Jefferson Davis, August 29, 1862, in Lee, *The Wartime Papers*, 266.

9. For the history of the battle, see Hennessy, *Return to Bull Run*. For a comprehensive look at the environmental influences on the campaign, especially regarding weather, food, and terrain, see Burns, "War and Nature in Northern Virginia."

10. McPherson, *Crossroads of Freedom*, 87.

11. McPherson, *Crossroads of Freedom*, 88–94, 98; Sears, *Landscape Turned Red*, 65–67; Ernst, *Too Afraid to Cry*, 39.

12. Lee to Davis, September 3, 1862, in Lee, *The Wartime Papers*, 293; Ernst, *Too Afraid to Cry*, 40.

13. Hunter, "A High Private's Account of the Battle of Sharpsburg," 507; Priest, *Before Antietam*, 44; Priest, *Antietam*, 3.

14. Bohannon, "Dirty, Ragged, and Ill-Provided For," 110–17; Hunter, "A High Private's Account of the Battle of Sharpsburg," 507.

15. Sears, *Landscape Turned Red*, 90–92, 112–13, 161; McPherson, *Crossroads of Freedom*, 113. For a fuller description of the Battle of South Mountain, see Sears, *Landscape Turned Red*, 117–49.

16. Cox, "The Battle at Antietam," 630–31; Alexander, "Destruction, Disease, and Death," 148–49; Ernst, *Too Afraid to Cry*, 12.

17. Bohannon, "Dirty, Ragged, and Ill-Provided For," 117; McPherson, *Crossroads of Freedom*, 138–46.

18. Bohannon, "Dirty, Ragged, and Ill-Provided For," 130–33.

19. Alexander, "Destruction, Disease, and Death," 151, 152, 155; "Population of the 100 Largest Urban Places"; Ernst, *Too Afraid to Cry*, 160, 185.

20. Van Buren, *Rules for Preserving the Health of the Soldier*, 5; Humphreys, *Marrow of Tragedy*, 98; Adams, *Living Hell*, 43.

21. Tao and Manci, "Estimating Manure Production."

22. Tarr, *The Search for the Ultimate Sink*, 114. Buffalo, with a population of 80,000 (just slightly larger than the Union encampment at Sharpsburg) experienced its last serious cholera epidemic in 1854. Installation of a sewer system in Chicago required that all the buildings be raised (sometimes as much as fourteen feet) above the level of Lake Michigan; by the 1880s, that remarkable engineering effort had all but eliminated cholera from the city.

23. Tarr, *The Search for the Ultimate Sink*, 113. Even in cities, residents relied on privies, outhouses, and cesspools before the advent of running water systems. Alexander, "Destruction, Disease, and Death," 156–58.

24. Ernst, *Too Afraid to Cry*, 194.

25. Entry dated May 23, 1862, in Stone, *Brokenburn*, 112–13; Steinberg, *Down to Earth*, 96; entries for April 12 through May 16, 1862, in Harris, *Piedmont Farmer*, 241–46; Thomas Cahill to Margaret Cahill, June 23, 1862, in Cahill, *The Greatest Trials I Ever Had*, 92.

26. Smith, *Starving the South*, 132; Noe, "The Drought That Changed the War"; entries dated June 1, May 28, and May 16, 1862, in Harris, *Piedmont Farmer*, 248, 250, 246.

27. Hilliard, *Hog Meat and Hoecake*, 40–49, 137–51; Smith, *Starving the South*, 31.

28. Entries dated July 2, September 13, 1862, in Harris, *Piedmont Farmer*, 252, 259; S. J. Westall to Zebulon Baird Vance, in Vance, *The Papers of Zebulon Baird Vance*, microfilm, reel 21; Smith, *Starving the South*, 132.

29. Stoll, *Larding the Lean Earth*, 120–22; Sutter, *Let Us Now Praise Famous Gullies*, 159; Mauldin, *Unredeemed Land*, 27–28.

30. Mauldin, *Unredeemed Land*, 25–26, 57–58; Stoll, *Larding the Lean Earth*, 128–29; Steinberg, *Down to Earth*, 87.

31. Crawford, *Ashe County's Civil War*, 87; S. J. Westall to Zebulon Baird Vance, January 17, 1864, in Vance, *The Papers of Zebulon Baird Vance*, microfilm, reel 21.

32. McPherson, *Battle Cry of Freedom*, 429–31; Leonidas L. Polk to wife, April 19, 1863, [typescript], L. Polk Denmark Collection, Private Collections, State Archives of North Carolina; Mary C. Williams and others, to Zebulon Vance, January 23, 1863, Governors Papers, State Archives of North Carolina; entry dated May 16, 1862, in Harris, *Piedmont Farmer*, 246.

33. Blair, *Virginia's Private War*, 39.

34. Smith, *Starving the South*, 20–22; Massey, *Ersatz in the Confederacy*, 63–64; Otto, *Southern Agriculture*, 23–30, 36; Silver, *Mount Mitchell and the Black Mountains*, 124. For more on the difficulties of acquiring food, see Cashin, *War Stuff*.

35. Smith, *Starving the South*, 14; Browning, *Shifting Loyalties*, 11, 55–80; Hilliard, *Hog Meat and Hoecake*, 101, 120–21, 220; U.S. Census Bureau, "1860 Census: Agriculture of the United States." Additionally, New Orleans, a city that had served as a major collection and distribution point for southern staples, had fallen to Federal forces on May 1, 1862. Over the course of the next year, Union forces gained control over much of the Mississippi River. Corn, meat, and other commodities that had once traveled the Mississippi to New Orleans now went to Union supply centers.

36. McPherson and Hogue, *Ordeal by Fire*, 238–39; Graebner, "Northern Diplomacy and European Neutrality," 66–68.

37. Entry dated March 17, 1863, in Stone, *Brokenburn*, 180–81; Poe, *True Tales of the South at War*, 59–60. Union cavalry soldiers participating in Benjamin Grierson's raid in Mississippi in April 1863 complained about the lack of coffee, and detested the assorted "secesh coffees" they encountered. Brown, *Grierson's Raid*, 137.

38. McCurry, *Confederate Reckoning*, 181, 210; Williams, "Bitterly Divided," 24. For more on how decreased food production exacerbated tensions on the home front, see also Cashin, *War Stuff*.

39. Blair, *Virginia's Private War*, 74; entry dated April 2, 1863, in Jones, *A Rebel War Clerk's Diary*, 183; McCurry, *Confederate Reckoning*, 193.

40. Todd, *Confederate Finance*, 166–68; Smith, *Starving the South*, 43; entry dated December 6, 1863, in Jones, *A Rebel War Clerk's Diary*, 315–16.

41. Todd, *Confederate Finance*, 141–42, 166–69.

42. Entries dated October 1, 1862, and July 17, 1863, and January 1, 1864, through March 20, 1865, in Jones, *A Rebel War Clerk's Diary*, 102, 243, 322–520.

43. Smith, *Starving the South*, 96–97; Silber and Stevens, *Yankee Correspondence*, 97; entry for July 1, 1863, in McSween, *Confederate Incognito*, 178; Zebulon Vance to James Seddon, December 21, 1863, U.S. War Department, *The War of the Rebellion*, ser. 4, 2:1061–62 (hereafter *War of the Rebellion*).

44. Sutherland, *A Savage Conflict*, 134, 155, 197; Fellman, *Inside War*, 23–26, 155–56. See also Stith, *Extreme Civil War*. For a more detailed study of the devastating effects of the Helena campaigns on the local food supply, see Schieffler, "Civil War in the Delta," 137–243.

45. Donahue, *The Great Meadow*, 25–29.

46. Stoll, *Larding the Lean Earth*, 81–96.

47. Smith, *Starving the South*, 3; McPherson and Hogue, *Ordeal by Fire*, 10; Atack and Bateman, *To Their Own Soil*, 224.

48. U.S. Census Bureau, "1860 Census: Population of the United States"; Smith, *Starving the South*, 73, 85–86; Steinberg, *Down to Earth*, 95. For more on northern farms and agrarian virtue during the war, see Dean, *An Agrarian Republic*.

49. Smith, *Starving the South*, 80–84.

50. McPherson and Hogue, *Ordeal by Fire*, 6; Smith, *Starving the South*, 73; Sutherland, *Seasons of War*, 296.

51. Robert E. Lee to Jefferson Davis, January 19, 1863, and Robert E. Lee to James Seddon, March 27, 1863, in Lee, *The Wartime Papers*, 392, 418; Sears, *Chancellorsville*, 36; Smith, *Starving the South*, 45.

52. Sears, *Chancellorsville*, 35–36; Wert, *General James Longstreet*, 233–38; James Longstreet to James Seddon, April 27, 1863, in *War of the Rebellion*, ser. 1, 18:1025.

53. Robert E. Lee to James Seddon, March 27, 1863, in Lee, *The Wartime Papers*, 418; entry dated April 20, 1863, in Jones, *A Rebel War Clerk's Diary*, 192; Furguson, *Chancellorsville, 1863*, 36–37; Sears, *Chancellorsville*, 36; Tally Simpson to Caroline Virginia Taliaferro Miller, April 25, 1863, in Simpson, *"Far, Far from Home,"* 218–19; entry for May 9, 1863, in Fisk, *Hard Marching Every Day*, 80.

54. Sears, *Chancellorsville*, 136–224. Lee did not urgently request that Longstreet return until April 29, too late for him to arrive in time to participate in the campaign. S. Cooper to Longstreet, April 29, 1863, in *War of the Rebellion*, ser. 1, vol. 25, pt. 2:757–58.

55. For comprehensive accounts of the battle, see Sears, *Chancellorsville*; and Furguson, *Chancellorsville, 1863*.

56. McPherson, *Battle Cry of Freedom*, 645. Mark Fiege persuasively argues that Lee invaded Pennsylvania primarily to provide food for his men and animals. See Fiege, "Gettysburg and the Organic Nature of the American Civil War."

57. Letter to father, July 20, 1863, in Hubard, *The Civil War Memoirs of a Virginia Cavalryman*, 120; Alexander, *Fighting for the Confederacy*, 229; Tally Simpson to Anna Tallulah Simpson, July 27, 1863, and Tally Simpson to Caroline Virginia Taliaferro Miller, June 28, 1863, in Simpson, *"Far, Far from Home,"* 261–62, 251.

58. John Bell Hood quoted in Guelzo, *Gettysburg*, 72; Tally Simpson to Anna Tallulah Simpson, July 27, 1863, and Tally Simpson to Caroline Virginia Taliaferro Miller, in Simpson, *"Far, Far from Home,"* 261–62.

59. Ballard, *The Civil War in Mississippi*, 35; Schneider, *Old Man River*, 284.

60. Morris, *Becoming Southern*, 103–16, 157–58, 183; Hoehling, *Vicksburg*, 2; "1860 Census: Population of the United States."

61. Ballard, *The Civil War in Mississippi*, 35–52; William Sherman to Ellen Ewing Sherman, May 25, 1863, in Sherman, *Sherman's Civil War*, 472; Morris, *Becoming Southern*, 157–58, 183.

62. Ballard, *The Civil War in Mississippi*, 35–105; Woodworth, *Nothing but Victory*, 244–54.

63. Woodworth, *Nothing but Victory*, 252–84; Grant, *Personal Memoirs*, 1:435. Unaware that Grant had turned back, Sherman launched an unsupported attack at Chickasaw Bayou in late December that failed, causing him to retreat as well.

64. Abstract from the Return of the Department of the Tennessee, April 30, 1863, in *War of the Rebellion*, ser. 1, vol. 24, pt. 3:250; William Sherman to Ellen Ewing Sherman, April 23, 1863, in Sherman, *Sherman's Civil War*, 455; Ballard, *The Civil War in Mississippi*, 139–47. For more on the environmental consequences of the Vicksburg campaign, see Brady, *War upon the Land*, 24–71. Nature altered the river years later, as it no longer follows the hairpin bend near Vicksburg.

65. Wheeler, *The Siege of Vicksburg*, 122, 162; William Sherman to Ellen Ewing Sherman, May 9, 1863, in Sherman, *Sherman's Civil War*, 470; Grant, *Personal Memoirs*, 1:435.

66. Leckie and Leckie, *Unlikely Warriors*, 84–99.

67. Leckie and Leckie, *Unlikely Warriors*, 84–99; Brown, *Grierson's Raid*, 39–40.

68. Smith, *The Smell of Battle*, 100–106; Wheeler, *The Siege of Vicksburg*, 135; Hoehling, *Vicksburg*, 201; Morris, *Becoming Southern*, 183.

69. Carlos Colby letter dated June 9, 1863, in Painter, "Bullets, Hardtack and Mud," 157. For a detailed campaign history, see Ballard, *Vicksburg*. In his last message to Pemberton,

Johnston said that he hoped to attack Grant's force on July 7 but that Pemberton would have to do most of the heavy work. Pemberton surrendered the next day. Johnston to Pemberton, July 3, 1863, in *War of the Rebellion*, ser. 1, vol. 24, pt. 3:987.

70. Smith, *The Smell of Battle*, 100–106; Hoehling, *Vicksburg*, 165; Wheeler, *The Siege of Vicksburg*, 180, 192, 194, 202.

71. Collingham, *The Taste of War*, 5–6; Hoehling, *Vicksburg*, 91, 99, 159, 130.

72. Smith, *The Smell of Battle*, 100–106; Wheeler, *The Siege of Vicksburg*, 205, 210; Ballard, *Vicksburg*, 382; Osborn, "A Tennessean at the Siege of Vicksburg, 367. For an examination of the problematic water resources, and the resulting health problems they caused for both armies, see Privette, "Fighting Johnnies, Fevers, and Mosquitoes."

73. Oldroyd, *A Soldier's Story*, 138; Hoehling, *Vicksburg*, 201, 121.

74. "Many Soldiers" to John C. Pemberton, June 28, 1863, in *War of the Rebellion*, ser. 1, 25:118–19; Hoehling, *Vicksburg*, 256.

75. Hoehling, *Vicksburg*, 287; Report of Stores on Hand at Commissary Depot, July 4, 1863, in *War of the Rebellion*, ser. 1, vol. 24, pt. 3:987; Woodworth, *Nothing but Victory*, 453; U.S. Census Bureau, "1860 Census: Agriculture of the United States."

76. Davis, *Besieged*, 35, 116, 311.

CHAPTER 4

1. Heiser, "The Widow and Her Farm."

2. Coco, *A Strange and Blighted Land*, 59–60; Heiser, "The Widow and Her Farm."

3. Coco, *A Strange and Blighted Land*, 75.

4. Coco, *A Strange and Blighted Land*, 83.

5. McClellan, *The Life and Campaigns of Major General J. E. B. Stuart*, 293–94; Eicher, *The Longest Night*, 493.

6. Greene, *Horses at Work*, 123, 127, 145; Jones, "The Influence of Horse Supply," 359.

7. Kennedy, *Agriculture of the United States in 1860*, cix; Jones, "The Influence of Horse Supply," 360; Greene, *Horses at Work*, 127–29, 134.

8. Phillips, "Writing Horses into American Civil War History," 167–74.

9. Morris, "Cavalry Horses in America," 160–61.

10. Morris, "Cavalry Horses in America," 159, 162; Ringwalt, "The Horse," 324, 325.

11. Ramsdell, "General Robert E. Lee's Horse Supply," 758; Phillips, "Writing Horses into American Civil War History," 170; Gerleman, "Unchronicled Heroes," 264, 303–5; Glatthaar, *Soldiering in the Army of Northern Virginia*, 33–43.

12. Ramsdell, "General Robert E. Lee's Horse Supply," 759, 770–72; Armistead, *Horses and Mules in the Civil War*, 95.

13. George Stoneman to Edwin Stanton, October 15, 1863, U.S. War Department, *The War of the Rebellion*, ser. 3, 3:885 (hereafter *War of the Rebellion*); Gates, *Agriculture and the Civil War*, 185; Starr, *The Union Cavalry in the Civil War*, 2:4–7; Phillips, "Writing Horses into American Civil War History," 179; Greene, *Horses at Work*, 142; M. C. Meigs to Edwin Stanton, November 3, 1864, *War of the Rebellion*, ser. 3, 4:889.

14. Wilson, *The Business of Civil War*, 140–47; Phillips, "Writing Horses into American Civil War History," 174.

15. Greene, *Horses at Work*, 145–46; Wilson, *The Business of Civil War*, 1; Gerleman, "Unchronicled Heroes," 370.

16. Larson, "The Horses of War"; McShane and Tarr, *The Horse in the City*, 3–4, 2.

17. Greene, *Horses at Work*, 135; Gerleman, "Unchronicled Heroes," 155–56.

18. Gerleman, "Unchronicled Heroes," 161–62; Gates, *Agriculture and the Civil War*, 133.

19. Lee to wife, February 8, 1863, and Lee to Jefferson Davis, April 16, 1863, in Lee, *The Wartime Papers*, 401, 435; Ramsdell, "General Robert E. Lee's Horse Supply," 760.

20. Zebulon Vance to James Seddon, January 22, 1863, in Zebulon B. Vance, Governors Papers, State Archives of North Carolina; Crawford, *Ashe County's Civil War*, 107.

21. Ramsdell, "General Robert E. Lee's Horse Supply," 765–67.

22. Greene, *Horses at Work*, 139. Spending on horseshoes tabulated from *War of the Rebellion*, ser. 3, 1:681, 2:788, 3:1120, 4:876, 5:252.

23. Wise, *The Long Arm of Lee*, 48–50.

24. Report of Brig. Gen. Rufus Ingalls, February 17, 1863, in *War of the Rebellion*, ser. 1, vol. 19, pt. 1:95.

25. Gerleman, "Unchronicled Heroes," 222–27; entry dated November 2 and 9, 1862, in Wainwright, *A Diary of Battle*, 121, 124; Robert E. Lee to George W. Randolph, November 7, 1862, in Lee, *The Wartime Papers*, 328. To make matters worse that fall, horses affected with greased heel also showed symptoms of "black tongue," possibly some sort of bacterial infection that caused swelling and discoloration of the mouth and tongue. Afflicted animals sometimes did not eat for four or five days. The lack of proper nutrition only made it more difficult to fend off greased heel. Gerleman, "Unchronicled Heroes," 226–27.

26. Gerleman, "Unchronicled Heroes," 227–28.

27. Sharrer, "The Great Glanders Epizootic," 83–84.

28. McShane and Tarr, *The Horse in the City*, 159; Sharrer, "The Great Glanders Epizootic," 91; Southern Memorial Association, "The Civil War Quartermaster's Glanders Stable."

29. Stoneman to Stanton, October 15, 1863, in *War of the Rebellion*, ser. 3, 3:886; Greene, *Horses at Work*, 142, 158; Ramsdell, "General Robert E. Lee's Horse Supply," 773. Spending on veterinary surgeons tabulated from *War of the Rebellion*, ser. 3, 1:681, 2:788, 3:1120, 4:876, 5:252.

30. Jones, "The Influence of Horse Supply," 369; Steinberg, *Down to Earth*, 92.

31. Jones, "The Influence of Horse Supply," 368; Greene, *Horses at Work*, 145.

32. Robert E. Lee to Jefferson Davis, August 24, 1863, in Lee, *The Wartime Papers*, 594; Griffith, "Grains for Horses and Their Characteristics."

33. Jno. Chambliss to A. R. Boteler, August 10, 1863, in *War of the Rebellion*, ser. 4, 2:720; Robert E. Lee to Jefferson Davis, July 24, 1863, August 24, 1863, in Lee, *The Wartime Papers*, 558; Jones, "The Influence of Horse Supply," 362, 365, 372–73, 594.

34. Steinberg, *Down to Earth*, 93; Powell, *The Chickamauga Campaign*, 196–98.

35. Jones, "The Influence of Horse Supply," 369–70; Link, *Atlanta*, 12; Hall, *Plowshares to Bayonets*, 23.

36. Eicher, *The Longest Night*, 590; Braxton Bragg to S. Cooper, December 28, 1863, *War of the Rebellion*, ser. 1, vol. 30, pt. 2:37.

37. Braxton Bragg to S. Cooper, December 28, 1863, *War of the Rebellion*, ser. 1, vol. 30, pt. 2:37; McDonough, *Chattanooga*, 48.

38. Greene, *Horses at Work*, 148–49; Armistead, *Horses and Mules in the Civil War*, 53.

39. McDonough, *Chattanooga*, 45–49; Armistead, *Horses and Mules in the Civil War*, 42.

40. "Visible Proofs."

41. Miller, "Historical Natural History," 228–30.

42. Bonhotal et al., "Horse Mortality."

43. Entry dated June 30, 1862, in Sneden, *Eye of the Storm*, 88.

44. Maryland Cooperative Extension, "Ask the Experts"; Armistead, *Horses and Mules in the Civil War*, 42; Coco, *A Strange and Blighted Land*, 60.

45. Sutherland, *Seasons of War*, 172; Dyer, *The Journals of a Civil War Surgeon*, 41; Greene, *Horses at Work*, 161.

46. Coco, *A Strange and Blighted Land*, 59–60.

47. "Food Habits of Feral Hogs"; Steinberg, *Down to Earth*, 188.

48. Steinberg, *Down to Earth*, 188; Wilson, "Pork," 88.

49. Bass, "How 'bout a Hand for the Hog," 305–8; Moore, *Confederate Commissary General*, 57.

50. Kirby, "The American Civil War."

51. Bass, "How 'bout a Hand for the Hog," 305–6; Walker, *The Statistics of the Wealth and Industry of the United States, 1870*, 87; Hilliard, *Hog Meat and Hoecake*, 42–43; Cronon, *Nature's Metropolis*, 226–47.

52. Bass, "How 'bout a Hand for the Hog," 304.

53. Hilliard, *Hog Meat and Hoecake*, 101–2; Cronon, *Nature's Metropolis*, 228–30; Smith, *Starving the South*, 30.

54. Moore, *Confederate Commissary General*, 57; Smith, *Starving the South*, 31.

55. Brady, *War upon the Land*, 11; Smith, *Starving the South*, 96–97; Grenville Dodge to Capt. S. Wait, May 5, 1863, *War of the Rebellion*, ser. 1, vol. 23, pt. 1:249.

56. Charles Rivers Ellet to Alfred W. Ellet, February 5, 1863, *War of the Rebellion*, ser. 1, vol. 24, pt. 1:338; John Adams to Henry Robinson, June 5, 1863, and William Sherman to U. S. Grant, July 14, 1863, *War of the Rebellion*, ser. 1, vol. 24, pt. 2:442, 526.

57. James Seddon to D. S. Donelson, February 14, 1863, H. Marshall to D. S. Donelson, April 18, 1863, John J. Walker to Braxton Bragg, February 23, 1863, *War of the Rebellion*, ser. 1, vol. 23, pt. 2:635, 777, 648–49.

58. Smith, *Starving the South*, 121–23; Potter, "Jefferson Davis and the Political Factors in Confederate Defeat," 97–98.

59. L. B. Northrop, Report on Subsistence [November 20, 1863], *War of the Rebellion*, ser. 4, 2:968–71; entry dated November 20, 1863, Jones, *A Rebel War Clerk's Diary*, 309; S. B. French to Northrop, December 8, 1863, in *War of the Rebellion*, ser. 4, 2:960.

60. Sharrer, *A Kind of Fate*, 18–19. U.S. Department of Agriculture, *Hog Cholera*, 9–11; Moore, *Confederate Commissary General*, 76.

61. Sharrer, *A Kind of Fate*, 19; Gates, *Agriculture in the Civil War*, 90; Coulter, "The Impact of the Civil War upon Pulaski County, Arkansas," 74; Mauldin, *Unredeemed Land*, 90.

62. Hall and Silver, "Nutrition and Feeding of the Calf-Cow Herd"; Stewart, "Whether Wast, Deodand, or Stray," 6–7, 17.

63. Kennedy, *Agriculture in the United States in 1860*, cviii–cix; Hilliard, *Hog Meat and Hoecake*, 129, 137; Morris, *Becoming Southern*, 25; Blevins, "Cattle Raising in Antebellum Alabama," 270–88; McMillen, "Beef," 26; Rubin, "The Limits of Agricultural Progress," 364–66.

64. Strom, *Making Catfish Bait Out of Government Boys*, 9–15; Altonen, "1851–1917, Cattle Drives and Texas Fever"; Haygood, "Cows, Ticks, and Disease," 551–64.

65. Gates, *The Farmer's Age*, 22–29; Stoll, *Larding the Lean Earth*, 81–96; Mauldin, *Unredeemed Land*, 15–27.

66. Indiana Division of Forestry, "Livestock Grazing in Woodlands"; Strom, *Making Catfish Bait Out of Government Boys*, 17.

67. Kennedy, *Agriculture in the United States in 1860*, cviii–cix; Henlein, *Cattle Kingdom in the Ohio Valley*, 111; Blevins, "Cattle Raising in Antebellum Alabama," 288.

68. McMillen, "Beef," 26; Hilliard, *Hog Meat and Hoecake*, 130.

69. Dan Tyler to Henry Halleck, June 24, 1863, Report of Richard S. Ewell, August 14, 1863, in *War of the Rebellion*, ser. 1, vol. 27, pt. 2:29, 443; Coco, *A Strange and Blighted Land*, 75. During the Confederate retreat from Gettysburg, Union cavalry managed to retake about 850 head of the purloined beeves and sent them straight to their own commissary bureaus for processing. John Buford to Lt. Col. C. Ross Smith, August 27, 1863, in *War of the Rebellion*, ser. 1, vol. 27, pt. 1:930.

70. Report of M. C. Meigs, August 1863, in *War of the Rebellion*, ser. 3, 3:602; Circular of Major P. W. White, November 2, 1863, in *War of the Rebellion*, ser. 1, vol. 28, pt. 2:471–73; Hurt, *Agriculture and the Confederacy*, 140; Smith, *Starving the South*, 115, 189.

71. Blair, *Virginia's Private War*, 39; Wise, *The Long Arm of Lee*, 47–48. Confederate soldiers also highly valued captured Union bridles due to the shortage of southern supply. Jones, "The Influence of Horse Supply," 366.

72. Report of R. E. Rodes, n.d. [1863], in *War of the Rebellion*, ser. 1, vol. 27, pt. 2:550; Robert E. Lee to Alexander Lawton, January 19, 1864, in Lee, *The Wartime Papers*, 653; Robert E. Lee to A. R. Lawton, January 30, 1864, in *War of the Rebellion*, ser. 1, 33:1131–32.

73. Steinberg, *Down to Earth*, 96–97; Guinn, "Exceptional Years," 36–37.

CHAPTER 5

1. Evans, *The 16th Mississippi Infantry*, 256; Krick, "An Insurmountable Barrier," 92–100.

2. Evans, *The 16th Mississippi Infantry*, 255; Krick, "An Insurmountable Barrier," 98; Report of Nathaniel Harris, in U.S. War Department, *The War of the Rebellion*, ser. 1, 36:1092 (hereafter *War of the Rebellion*); Matter, *If It Takes All Summer*, 211, 260. The number of men in Harris's brigade is difficult to ascertain. As Robert K. Krick analyzed, two extant sources—one by Lee's aide Charles Venable and another by Harris—claim 800 and 1,600 men, respectively, but both were written many years after the battle. Krick split the difference and estimated 1,200 men present. Other historians accept Venable's number of 800, and we have chosen to do so as well.

3. Krick, "An Insurmountable Barrier," 101; Report of Nathaniel Harris, in *War of the Rebellion*, ser. 1, 36:1092; Evans, *16th Mississippi Infantry*, 258; Matter, *If It Takes All Summer*, 211, 260.

4. Rhea, *The Battles for Spotsylvania*, 311; Krick, "An Insurmountable Barrier," 107, 113; Porter, *Campaigning with Grant*, 111.

5. Rhea, *The Battles for Spotsylvania*, 311–12; Krick, "An Insurmountable Barrier," 113; McPherson, *Battle Cry of Freedom*, 731; Pfanz, "Burying the Dead at Spotsylvania."

6. Letter dated May [n.d.], 1864, in Fisk, *Hard Marching Every Day*, 221; Reardon, "A Hard Road to Travel," 170.

7. Hacker, "A Census-Based Count of the Civil War Dead," 307; Faust, *This Republic*

of Suffering, xi, xiii. According to one scholar, the "average person saw two to three times (and sometimes upward of six to seven times) as many people die per year than we do today." Marshall, "The Great Exaggeration," 4–5.

8. "Decomposition"; Joseph et al., "The Use of Insects in Forensic Investigations"; Costandi, "The Smell of Death"; Costandi, "Life after Death."

9. Coco, *A Strange and Blighted Land*, 89; Williams, *A People's History of the Civil War*, 228; Adams, *Living Hell*, 102.

10. Morgan, "Infectious Disease Risks from Dead Bodies."

11. Humphreys, *Marrow of Tragedy*, 79.

12. Costandi, "The Smell of Death"; "Decomposition"; Steiner, *Disease and the Civil War*, 135.

13. Hard, *History of the Eighth Cavalry Regiment*, 133.

14. Faust, *This Republic of Suffering*, 62; Daniel Holt to wife, September 16, 1862, in Holt, *A Surgeon's Civil War*, 21; McPherson, *Battle Cry of Freedom*, 731.

15. Coco, *A Strange and Blighted Land*, 90, 91.

16. Coco, *A Strange and Blighted Land*, 83, 120; Groeling, *The Aftermath of Battle*, 79–80.

17. Halleck to Grant, January 1864, in *War of the Rebellion*, ser. 1, vol. 32, pt. 2:122; McPherson, *Battle Cry of Freedom*, 722–25; Johnson, *Red River Campaign*, 80–100, 278–79; Engle, *Yankee Dutchman*, 174–95.

18. Foote, *The Civil War*, 3:111–12; Cimprich, *Fort Pillow*, 80–81.

19. Castel, *Decision in the West*.

20. Rhea, *The Battle of the Wilderness*, 125–53; Eicher, *The Longest Night*, 671; Young, *Lee's Army during the Overland Campaign*, 235; McPherson, *Battle Cry of Freedom*, 731.

21. Trudeau, *Bloody Roads South*, 213; Young, *Lee's Army during the Overland Campaign*, 236; McPherson, *Battle Cry of Freedom*, 731; Rhea, *Cold Harbor*, 319, 333, 362; Grant, *Personal Memoirs*, 2:276; Bonekemper, *A Victor, Not a Butcher*, 311.

22. Entry dated June 18, 1864, in Wainwright, *A Diary of Battle*, 425; Bonekemper, *A Victor, Not a Butcher*, 313.

23. Levin, "The Devil Himself Could Not Have Checked Them," 271, 272; Trudeau, *The Last Citadel*, 109–27.

24. Downs, *Sick from Freedom*, 18–41; Taylor, *Embattled Freedom*, 106–39; Trudeau, *The Last Citadel*, 125; Levin, "The Devil Himself Could Not Have Checked Them," 273.

25. Casualty figures can be found in Eicher, *The Longest Night*, 296, 334, 363, 405, 671, 679, 686.

26. Faust, *This Republic of Suffering*, 90–95.

27. Schmidt, "Six Feet Under."

28. Faust, *This Republic of Suffering*, 91.

29. Faust, *This Republic of Suffering*, 91; Schmidt, "Six Feet Under"; Gagnon, "Death and Mourning in the Civil War Era."

30. "Improved Burial Case," 136; Metcalf, "Inside the Patent Office's Cabinet"; Schmidt, "Six Feet Under."

31. Faust, *This Republic of Suffering*, 94; Lee, "The Undertaker's Role."

32. Faust, *This Republic of Suffering*, 94–96.

33. Gillet, *The Army Medical Department*, 153–54.

34. Freemon, *Gangrene and Glory*, 232; Earley, *The Second United States Sharpshooters*, 26; Fales, "M-16," 131.

35. Cunningham, *Doctors in Gray*, 5; Adams, *Living Hell*, 85–86; Gillet, *The Army Medical Department*, 166.

36. Byrne, "Divine Intervention via a Microbe."

37. Soniak, "Why Some Civil War Soldiers Glowed in the Dark."

38. Abbot, "A Newspaper for Injured Civil War Vets"; Bollet, *Civil War Medicine*, 212.

39. Kruglikova and Chernysh, "Surgical Maggots," 667–68; Chernin, "Surgical Maggots," 1144–45.

40. King, "Maggots: Friend or Foe?"; Kerr, *Civil War Surgeon*, 186–88; Bollet, *Civil War Medicine*, 212; "John Forney Zacharias," 748.

41. Kruglikova and Chernysh, "Surgical Maggots," 672.

42. Bollet, *Civil War Medicine*, 212.

43. McGaugh, *Surgeon in Blue*, 75–97.

44. McGaugh, *Surgeon in Blue*, 75–97; Adams, *Living Hell*, 86, 90; Humphreys, *Marrow of Tragedy*, 241–42; Bollet, *Civil War Medicine*, 186–88.

45. Freemon, *Gangrene and Glory*, 171–74; Bollet, *Civil War Medicine*, 130–31.

46. Bollet, *Civil War Medicine*, 198–200; Turkington and Ashby, *The Encyclopedia of Infectious Diseases*, 126.

47. Bollet, *Civil War Medicine*, 152–53; Nelson, *Ruin Nation*, 160–227. For more on amputees, see also Wegner, "Phantom Pain"; Figg and Farrell-Beck, "Amputation in the Civil War."

48. Williams, *A People's History of the Civil War*, 234; Morris, *The Better Angel*, 93; Browning, "All for One Charge," 29.

49. Bollet, *Civil War Medicine*, 4, 217; Nelson and Sheriff, *A People at War*, 122.

50. Gardner, "Prisoners of War," 3:1572.

51. Gray, "Elmira," 322–38; Meyer, "Elmira Prison," 2:648–49.

52. Davis, "Andersonville Prison"; Kellogg, *Life and Death in Rebel Prisons*, 56.

53. "Nature & Science"; Jordan, *Marching Home*, 136; Adams, *Living Hell*, 179; entries dated June 17, 1864, and June 26, 1864, in Hitchcock, *"Death Does Seem to Have All He Can Attend To,"* 184, 187, 201; Ruhlman, *Captain Henry Wirz*, 148.

54. Nelson and Sheriff, *A People at War*, 228; Jordan, *Marching Home*, 138.

55. Bollet, *Civil War Medicine*, 157.

56. Padilla, "Army of 'Cripples,'" 72–75.

57. Marten, *Sing Not War*, 77, 123.

58. Marten, *Sing Not War*, 200, 203, 214, 217; Skocpol, "America's First Social Security System," 85–86; Adams, *Living Hell*, 205; Jordan, *Marching Home*, 152.

59. *The Seventh Census of the United States . . . 1850*, lxxxii; *Agriculture of the United States in 1860*, vii, xxix, xlvi; *The Statistics of the Wealth and Industry of the United States . . . 1870*, 81, 83.

CHAPTER 6

1. Whisonant, *Arming the Confederacy*, 99, 102.

2. Whisonant, *Arming the Confederacy*, 99–100.

3. Mosgrove, *Kentucky Cavaliers in Dixie*, 201; Whisonant, *Arming the Confederacy*, 103–5. Much of our description of the battle derives from our visit to the Museum of the Middle Appalachians in Saltville, Virginia, as well as our walking the ground where the fighting took place.

4. Mosgrove, *Kentucky Cavaliers in Dixie*, 203.

5. Mosgrove, *Kentucky Cavaliers in Dixie*, 207; Whisonant, *Arming the Confederacy*, 106–7.

6. Whisonant, *Arming the Confederacy*, 88–94.

7. Even so, the huge cache of salt might have remained out of the reach of people had it not been for another geologic quirk: a large crack in the earth's crust—known to geologists as a fault—that extends some 450 miles from Virginia into Alabama. Ancient movement along this fracture, now called the Saltville Fault, forced older formations up and over more recent deposits and pushed rocks that might otherwise have remained buried toward the surface, including some from the Maccrady Formation. Whisonant, *Arming the Confederacy*, 95.

8. Worrall, "When, How Did the First Americans Arrive?"; "General Burbridge's Raid in Southwestern Virginia," 21; Lonn, *Salt as a Factor in the Confederacy*, 68.

9. Allison, Glanville, and Haynes, "Saltville during the Civil War," 73–76; Mauldin, *Unredeemed Land*, 64; Lonn, *Salt as a Factor in the Confederacy*, 120.

10. Whisonant, *Arming the Confederacy*, 23–34, 102.

11. Allison, Glanville, and Haynes, "Saltville during the Civil War," 74–75; Lonn, *Salt as a Factor in the Confederacy*, 121. The hardwood forests around Saltville consisted primarily of various oaks, hickories, and American chestnut. Cordwood cut from oaks and hickories burned hotter than chestnut, but chestnut was probably easier to come by since it grew faster. The authors thank Brian Ledford, the area forester for Smyth County, Virginia, for the Virginia Department of Forestry, who provided information on the types of trees and their growth rates.

12. Lonn, *Salt as a Factor in the Confederacy*, 59.

13. Mosgrove, *Kentucky Cavaliers in Dixie*, 203.

14. Allison, Glanville, and Haynes, "Saltville during the Civil War," 82; Whisonant, *Arming the Confederacy*, 51. Over the course of the war, the Union tried to limit Confederate access to other minerals that lay beneath the ground of southwestern Virginia. Iron, coal, and lead mines became targets for northern troops, especially during the last two years of the war. On more than one occasion, enemy soldiers raided Virginia caves in search of saltpeter, a substance critical in the manufacture of gunpowder. In the caves, a peculiar combination of bacterial action, moving water, and stable temperatures created nitrate-laden soil that could be easily refined into the saltpeter.

15. For more on Jackson's Shenandoah Valley campaign, see Cozzens, *Shenandoah, 1862*.

16. Ehlen and Whisonant, "Military Geology of Antietam Battlefield"; Thompson, "With Burnside at Antietam," 660. See Brady and Silver, "Nature as Material Culture."

17. Cozzens, *No Better Place to Die*, 122; National Park Service, Geologic Resources Division, *Stones River National Battlefield*, 15–17.

18. Cozzens, *No Better Place to Die*, 125–26; Eicher, *The Longest Night*, 428; Jones, *Civil War Command*, 108.

19. Brown, *Geology and the Gettysburg Campaign*, 5–7.

20. Sears, *Gettysburg*, 269–83.

21. Pfanz, *Gettysburg*, 205–34; Sears, *Gettysburg*, 359–468.

22. Zax, "Civil War Geology."

23. Adams, "Iron from the Wilderness," 7–9.

24. Adams, "Iron from the Wilderness," 44–48.

25. Kneipp, "The Wilderness as Wilderness," 2320–28.

26. Rhea, *Battle of the Wilderness*, 125.

27. Rhea, *Battle of the Wilderness*, 142, 150, 153.

28. Rhea, *Battle of the Wilderness*, 208.

29. Krick, "Lee to the Rear"; Rhea, *Battle of the Wilderness*, 351–66.

30. Rhea, *Battle of the Wilderness*, 366–403.

31. Kennett, *Marching through Georgia*, 112–14; Davis, "A Very Barbarous Mode of Carrying on War," 57; Borden, *Thirsty City*, 6–10.

32. Kennett, *Marching through Georgia*, 125; Wortman, *The Bonfire*, 283, 292; Davis, "A Very Barbarous Mode of Carrying on War," 57, 79–80; Carter, *The Siege of Atlanta*, 284, 392; letter dated October 30, 1864, in Cram, *Soldiering with Sherman*, 147; Link, *Atlanta*, 29; Leigh, "Who Burned Atlanta?"

33. Sherman, *Memoirs*, 493–95; Davis, "A Very Barbarous Mode of Carrying on War," 86–88.

34. Brady, "The Wilderness of War," 176.

35. Link, *Atlanta*, 48; Elmore, "Hydrology and Residential Segregation," 45–47; Garrett, *Atlanta and Environs*, 2:1.

36. Elmore, "Hydrology and Residential Segregation," 39–40, 58–59; Link, *Atlanta*, 194.

37. Letter dated October 30, 1864, in Cram, *Soldiering with Sherman*, 148; Davis, "A Very Barbarous Mode of Carrying on War," 85; Elmore, "Hydrology and Residential Segregation," 41; Mauldin, *Unredeemed Land*, 52.

38. Nelson, *Ruin Nation*, 114–16, 139; Gates, *Agriculture and the Civil War*, 138; Mauldin, *Unredeemed Land*, 60.

39. Mauldin, *Unredeemed Land*, 27; Dyer, *The Journals of a Civil War Surgeon*, 121; letter dated January 5, 1865, in Cram, *Soldiering with Sherman*, 154.

40. Nelson, *Ruin Nation*, 118; McSween, *Confederate Incognito*, 91.

41. Dodge, *On Campaign with the Army of the Potomac*, 218; Powell, "Seven Year Locusts."

42. Entry dated July 4, 1863, in Dodge, *On Campaign with the Army of the Potomac*, 306; letter dated May 3, 1864, in Cram, *Soldiering with Sherman*, 95; Mauldin, "Unredeemed Land," 148.

43. Nelson, *Ruin Nation*, 152.

44. Mauldin, *Unredeemed Land*, 51–52; Nelson, *Ruin Nation*, 151–52.

45. U.S. Census Bureau, "1860 Census: Agriculture of the United States"; U.S. Census Bureau, "1870 Census."

46. U.S. Census Bureau, "1860 Census: Agriculture of the United States"; U.S. Census Bureau, "1870 Census."

47. Swanson, "War Is Hell, So Have a Chew," 169–73.

48. Swanson, "War Is Hell, So Have a Chew," 173–74.

49. Mauldin, "Freedom, Economic Autonomy, and Ecological Change," 403–7.

50. Mauldin, "Freedom, Economic Autonomy, and Ecological Change," 412, 415; Stewart, *What Nature Suffers to Groe*.

51. DeGruccio, "Letting the War Slip through Our Hands," 29.

52. National Park Service, "Cultural Landscapes Inventory," 2.

53. Rable, *Fredericksburg! Fredericksburg*, 218–70; Mink, "Can't See the Battlefield for the Trees."

54. The information for the above paragraphs comes from Brady and Silver, "Nature as Material Culture," 53–74.

55. Neil, "'Friends' Take Active Role in Park Support," 1, 12; National Park Service, "Frequently Asked Questions"; Black, "The Copse at Gettysburg," 307–9.

EPILOGUE

1. Varon, *Appomattox*, 89.
2. Marvel, *A Place Called Appomattox*, 246–47; Varon, *Appomattox*, 90.
3. Humphreys, *Marrow of Tragedy*, 92; Downs, *Sick from Freedom*, 11–17, 21–24, 95–119.
4. Lowry, *The Story the Soldiers Wouldn't Tell*, 108.
5. Devine, *Learning from the Wounded*, 10.
6. DeBruler, "Remarks on Chronic Diarrhea," 3:42; Jones, "Opium Slavery," 4; Courtwright, "Opiate Addiction," 101–11; Adams, *Living Hell*, 199.
7. Jordan, *Marching Home*, 127. See also Handley-Cousins, *Bodies in Blue*.
8. Raines, *Getting the Message Through*, 3–31, 44–45.
9. Myer, *Annual Report of the Chief Signal Officer*, 16; Raines, *Getting the Message Through*, 45–47.
10. Myer, *Annual Report of the Chief Signal Officer*, 16; National Park Service, "Ulysses S. Grant Timeline."
11. Sharrer, *A Kind of Fate*, 17.
12. *Report on the Statistics of Agriculture . . . 1890*, 86–115. The 1890 agricultural census listed the agricultural totals for the previous four decades for comparison; Sharrer, *A Kind of Fate*, 10.
13. Guinn, "Exceptional Years," 36–37; Reyes, "Dry Years of 1862–65 Changed O.C. Life"; Wills, *Conservation Fallout*, 29–30; Kennedy, *Agriculture in the United States in 1860*, cviii; Walker, *The Statistics of the Wealth and Industry of the United States, 1870*, 87; Walker, *Reports on the Production of Agriculture . . . 1880*, 5.
14. *Report on the Statistics of Agriculture . . . 1890*, 86–115.
15. South Carolina's hog population fell from 965,779 in 1860 to a mere 150,000 animals by 1865, roughly an 85 percent reduction, with much of the state "utterly stripped of its stock." By 1870, the number of pigs in South Carolina had doubled to nearly 306,000, but even then the state had only about one-third as many hogs as before the war. We applied South Carolina's recovery between 1865 and 1870 as a method to estimate the losses in the other southern states. Bass, "How 'bout a Hand for the Hog," 308; *Report on the Statistics of Agriculture . . . 1890*, 86–115; *Twelfth Census of the United States*, ccxx–ccxxi. For more on modern factory hog farms, see Estabrook, *Pig Tales*.
16. Gates, *Agriculture and the Civil War*, 177, 183; Walker, *The Statistics of the Wealth and Industry of the United States, 1870*, 82, 87; Kirby, "The American Civil War," part 5.
17. Sharrer, "The Great Glanders Epizootic," 95; Beirer, *A Short History of Veterinary Medicine*, 40–41.
18. Sharrer, "The Great Glanders Epizootic," 96.
19. Hacker, "A Census-Based Count of the Civil War Dead," 307; "Civil War Casualties"; Hacker, Hilde, and Jones, "The Effect of the Civil War on Southern Marriage Patterns," 39–70.
20. *Report on the Statistics of Agriculture . . . 1890*, 86–115. The 1890 agricultural census listed the agricultural totals for the previous four decades for comparison.
21. *Report on the Statistics of Agriculture . . . 1890*, 86–115.

22. Kirby, "The American Civil War," part 4; *Report on the Statistics of Agriculture . . . 1890*, 86–115.

23. Hupy, "The Environmental Footprint of War," 414–17; Nelson, *Ruin Nation*, 152–54; Pearson, *Scarred Landscapes*, 40–47; Kirby, "The American Civil War," part 4.

24. Mauldin, "Freedom, Economic Autonomy, and Ecological Change," 403–7; Kirby, "The American Civil War," part 5.

25. Swanson, "War Is Hell, So Have a Chew," 181; Kirby, "The American Civil War," part 5.

26. Sachs, "Stumps in the Wilderness," 96.

27. Duncan, *Seed of the Future*, 67; Runte, *National Parks*, 30. See also Dean, *An Agrarian Republic*; Brady, *War upon the Land*.

28. Olmsted, *Yosemite and the Mariposa Grove*; Diamant, "Lincoln, Olmsted, and Yosemite," 12–13; Brady, *War upon the Land*, 136–37; Sutter, foreword, xiii–xiv.

29. Brady, *War upon the Land*, 137–39; for a synopsis of Transcendentalist thinking and preservation of nature, see Runte, *National Parks*, 11–32.

30. Runte, *National Parks*, 46; "Famous Quotes Concerning the National Parks"; Sutter, foreword, xiii–xiv; Gettysburg National Military Park, "History and Culture."

BIBLIOGRAPHY

MANUSCRIPTS

Chapel Hill, North Carolina
    Southern Historical Collection, Louis Round Wilson
    Library, University of North Carolina
        George Loyall Gordon Papers
Raleigh, North Carolina
    State Archives of North Carolina
        Governors Papers
        L. Polk Denmark Collection
Washington, D.C.
    National Archives
        Federal Pension Application Files, Records of the
        Veterans Administration, Record Group 15
Worcester, Massachusetts
    American Antiquarian Society
        Civil War Collection

PUBLISHED PRIMARY SOURCES

Agriculture of the United States in 1860: Compiled from the Original Returns of the Eighth Census. Washington, DC: U.S. Government Printing Office, 1864. http://usda.mannlib.cornell.edu/usda/AgCensusImages/1860/1860b-02.pdf.

Alexander, Edward Porter. *Fighting for the Confederacy: The Personal Recollections of General Edward Porter Alexander.* Edited by Gary W. Gallagher. Chapel Hill: University of North Carolina Press, 1989.

Barnes, Joseph K., comp. *The Medical and Surgical History of the War of the Rebellion (1861–1865), Prepared in Accordance with the Acts of Congress.* 6 vols. Washington, DC: U.S. Government Printing Office, 1870.

Baxter, J. H. *Statistics, Medical and Anthropological, of the Provost-Marshal-General's Bureau, Derived from Records of the Examination for Military Service in the Armies of the United States during the Late War of the Rebellion,* vol. 1. Washington, DC: U.S. Government Printing Office, 1875.

Berkeley, Henry Robinson. *Four Years in the Confederate Artillery: The Diary of Private Henry Robinson Berkeley*. Edited by William H. Runge. Chapel Hill: University of North Carolina Press, 1961.

Brewer, William H. *Up and Down California in 1860–1864: The Journal of William H. Brewer*. 4th ed. Edited by Francis P. Farquhar. Berkeley: University of California Press, 2003.

Cahill, Thomas. *The Greatest Trials I Ever Had: The Civil War Letters of Margaret and Thomas Cahill*. Edited by Ryan W. Keating. Athens: University of Georgia Press, 2017.

Callaway, Joshua K. *The Civil War Letters of Joshua K. Callaway*. Edited by Judith Lee Hallock. Athens: University of Georgia Press, 1997.

Clough, Phil P. J. "Battle of Valverde." In *Civil War in the Southwest: Recollections of the Sibley Brigade*, edited by Jerry D. Thompson, 44–49. College Station: Texas A&M University Press, 1991.

Cox, Jacob D. "The Battle at Antietam." In *North to Antietam: Battles and Leaders of the Civil War*, vol. 2, edited by Robert Underwood Johnson and Clarence Clough Buel, 630–59. New York: Castle, 1956.

Cram, George Franklin. *Soldiering with Sherman: Civil War Letters of George Franklin Cram*. Edited by Jennifer Cain Bohrnstedt. Dekalb: Northern Illinois University Press, 2000.

Davidson, William. "Battle of Valverde." In *Civil War in the Southwest: Recollections of the Sibley Brigade*, edited by Jerry D. Thompson, 27–43. College Station: Texas A&M University Press, 1991.

———. "Dona Ana to San Antonio." In Thompson, *Civil War in the Southwest*, 127–32.

———. "From Socorro to Glorieta." In Thompson, *Civil War in the Southwest*, 78–86.

DeBruler, James P. "Remarks on Chronic Diarrhea." In *The Medical and Surgical History of the War of the Rebellion (1861–1865), Prepared in Accordance with the Acts of Congress*. 6 vols. Compiled by Joseph K. Barnes, 3:42. Washington, DC: U.S. Government Printing Office, 1870.

Dodge, Theodore Ayrault. *On Campaign with the Army of the Potomac: The Civil War Journal of Theodore Ayrault Dodge*. Edited by Stephen W. Sears. New York: Cooper Square, 2001.

Duncan, Captain Louis C. "The Campaign of Fredericksburg, December 1862." *Military Surgeon* 33 (July 1913): 1–40.

Dyer, J. Franklin. *The Journals of a Civil War Surgeon*. Edited by Michael Chesson. Lincoln: University of Nebraska Press, 2003.

Evans, Robert G., ed. *The 16th Mississippi Infantry: Civil War Letters and Reminiscences*. Jackson: University Press of Mississippi, 2002.

"The Feeling in Edgecombe." *Tarborough Southerner*. January 26, 1861.

Fisk, Wilbur. *Hard Marching Every Day: The Civil War Letters of Private Wilbur Fisk, 1861–1865*. Edited by Emil and Ruth Rosenblatt. Lawrence: University Press of Kansas, 1922.

"General Burbridge's Raid in Southwestern Virginia." *Harper's Weekly* 14 (January 1865): 21.

Grant, Ulysses Simpson. *Personal Memoirs of U. S. Grant*. 2 vols. New York: Charles L. Webster, 1894.

"The Great Flood in California: Great Destruction of Property, Damage $10,000,000." *New York Times*, January 21, 1862.

Hard, Abner. *History of the Eighth Cavalry Regiment Illinois Volunteers during the Great Rebellion.* Aurora, IL: n.p., 1868.

Harris, David Golightly. *Piedmont Farmer: The Journals of David Golightly Harris, 1855–1870.* Edited by Philip N. Racine. Knoxville: University of Tennessee Press, 1990.

Hitchcock, George A. *"Death Does Seem to Have All He Can Attend To": The Civil War Diary of an Andersonville Survivor.* Edited by Ronald G. Watson. Jefferson, NC: McFarland, 2014.

Holt, Daniel M. *A Surgeon's Civil War: The Letters and Diary of Daniel M. Holt, M.D.* Edited by James M. Greiner, Janet L. Coryell, and James R. Smither. Kent, OH: Kent State University Press, 1994.

Hubard, Robert T., Jr. *The Civil War Memoirs of a Virginia Cavalryman.* Edited by Thomas P. Nanzig. Tuscaloosa: University of Alabama Press, 2007.

Hunter, Alexander. "A High Private's Account of the Battle of Sharpsburg." *Southern Historical Society Papers* 10 (October and November 1882): 503–12.

"Improved Burial Case." *Scientific American* 8, no. 9 (February 28, 1863): 136.

Johnson, Robert Underwood, and Clarence Clough Buel. *Battles and Leaders of the Civil War.* 4 vols. New York: Century, 1887–88.

Jones, John B. *A Rebel War Clerk's Diary.* Edited by Earl Schenck Miers. New York: A. S. Barnes, 1961.

Kellogg, Robert H. *Life and Death in Rebel Prisons.* Hartford, CT: L. Stebbins, 1865.

Kennedy, Joseph C. G., comp. *Agriculture of the United States in 1860.* Washington, DC: U.S. Government Printing Office, 1864. http://www.agcensus.usda.gov/Publications/Historical_Publications/1860/1860b-03.pdf.

Knox, Thomas W. *Camp-Fire and Cotton Field: Southern Adventure in Time of War, Life with the Union Armies, and Residence on a Louisiana Plantation.* Cincinnati: Jones Bros., 1865.

Lee, Robert E. *The Wartime Papers of Robert E. Lee.* Edited by Clifford Dowdey and Louis H. Manarin. New York: Da Capo, 1961.

McClellan, George B. *The Civil War Papers of George B. McClellan: Selected Correspondence, 1860–1865.* Edited by Stephen W. Sears. New York: Da Capo, 1992.

McClellan, Henry B. *The Life and Campaigns of Major General J. E. B. Stuart, Commander of the Cavalry of the Army of Northern Virginia.* Boston: Houghton, Mifflin, 1885.

McLaws, Lafayette. *A Soldier's General: The Civil War Letters of Major General Lafayette McLaws.* Edited by John C. Oeffinger. Chapel Hill: University of North Carolina Press, 2002.

McSween, Murdoch John. *Confederate Incognito: The Civil War Reports of "Long Grabs," a.k.a. Murdoch John McSween, 26th and 35th North Carolina Infantry.* Edited by E. B. Munson. Jefferson, NC: McFarland, 2013.

Morgan, W. H. *Personal Reminiscences of the War of 1861–65.* Lynchburg, VA: J. P. Bell, 1911.

Morris, Frances. "Cavalry Horses in America." In *Report of the Commissioner of Agriculture for the Year 1863,* 159–75. Washington, DC: U.S. Government Printing Office, 1863.

Mosgrove, George D. *Kentucky Cavaliers in Dixie: Reminiscences of a Confederate Cavalryman.* Edited by Bell Irvin Wiley. Jackson, TN: McCowart-Mercer, 1957.

Myer, Albert J. *Annual Report of the Chief Signal Officer to the Secretary of War.* Washington, DC: U.S. Government Printing Office, 1870.

National Park Service. "Ulysses S. Grant Timeline." National Park Service. https://www.nps.gov/ulsg/learn/historyculture/ulysses-s-grant-timeline.htm. Accessed November 2, 2018.

New York State Military Museum and Veterans Research Center. "35th Regiment New York Volunteer Infantry, Civil War Newspaper Clippings." NYS Division of Military and Naval Affairs. https://dmna.ny.gov/historic/reghist/civil/infantry/35thInf/35thInfCWN.htm. Accessed September 11, 2015.

Oldroyd, Osborn H. *A Soldier's Story of the Siege of Vicksburg*. Springfield, IL: Osborn Oldroyd, 1885.

Olmsted, Frederick Law. *Yosemite and the Mariposa Grove: A Preliminary Report, 1865*. http://www.yosemite.ca.us/library/olmsted/report.html.

Osborn, George C. "A Tennessean at the Siege of Vicksburg: The Diary of Samuel Alexander Ramsey Swan, May–July, 1863." *Tennessee Historical Quarterly* 14 (December 1955): 353–72.

Painter, John S., ed. "Bullets, Hardtack and Mud: A Soldier's View of the Vicksburg Campaign, from the Letters of Carlos W. Colby." *Journal of the West* 4 (April 1965): 129–68.

Porter, Horace. *Campaigning with Grant*. New York: Century, 1906.

Report on the Statistics of Agriculture in the United States at the Eleventh Census, 1890. Washington, DC: U.S. Government Printing Office, 1895. http://agcensus.mannlib.cornell.edu/AgCensus/censusParts.do?year=1890.

Ringwalt, Colonel Samuel. "The Horse: From Practical Experience in the Army." *Report of the Commissioner of Agriculture for the Year 1866*, 321–34. Washington, DC: U.S. Government Printing Office, 1867.

Robillard, Douglas, ed. *The Poems of Herman Melville*. Kent, OH: Kent State University Press, 1976.

Schmidgall, Gary, ed. *Intimate with Walt: Selections from Whitman's Conversations with Horace Traubel, 1888–1892*. Iowa City: University of Iowa Press, 2001.

The Seventh Census of the United States, 1850. Washington, DC: Robert Armstrong, 1853. http://usda.mannlib.cornell.edu/usda/AgCensusImages/1850/1850a-03.pdf.

Sherman, William Tecumseh. *Memoirs of General W. T. Sherman*. Edited by Michael Fellman. New York: Penguin, 2000.

———. *Sherman's Civil War: Selected Correspondence of William T. Sherman, 1860–1865*. Edited by Brooks D. Simpson and Jean V. Berlin. Chapel Hill: University of North Carolina Press, 1999.

Silber, Nina, and Mary Beth Stevens, eds. *Yankee Correspondence: Civil War Letters between New England Soldiers and the Home Front*. Charlottesville: University of Virginia Press, 1996.

Simpson, Richard W. *"Far, Far from Home": The Wartime Letters of Dick and Tally Simpson, Third South Carolina Volunteers*. Edited by Guy R. Everson and Edward H. Simpson. New York: Oxford University Press, 1994.

Sneden, Robert Knox. *Eye of the Storm: A Civil War Odyssey*. Edited by Charles F. Bryan Jr. and Nelson D. Lankford. New York: Free Press, 2000.

"Southern Rights Meeting." *Tarborough Southerner*. March 2, 1861.

The Statistics of the Wealth and Industry of the United States . . . From the Original Returns of the Ninth Census, June 1, 1870. Washington, DC: U.S. Government Printing Office, 1872. http://usda.mannlib.cornell.edu/usda/AgCensusImages/1870/1870c-02.pdf.

Stone, Kate. *Brokenburn: The Journal of Kate Stone, 1861–1868*. Edited by John Q. Anderson. Baton Rouge: Louisiana State University Press, 1972.

Taylor, William H. "Some Experiences of a Confederate Assistant Surgeon." *Transactions of the College of Physicians of Philadelphia*, 3rd ser., vol. 28, 91–121. Philadelphia: Dornan, 1906.

Thompson, David L. "With Burnside at Antietam." In *Battles and Leaders of the Civil War*, vol. 2. Edited by Robert Underwood Johnson and Clarence Clough Buel, 660–62. New York: Century, 1887.

Trobriand, Régis de. *Four Years with the Army of the Potomac*. Translated by George K. Dauchy. Boston: Ticknor, 1889.

Twelfth Census of the United States, Taken in the Year 1900: Agriculture, part 1, *Farms, Live Stock, and Animal Products*. Washington, DC: U.S. Census Office, 1902. http://agcensus.mannlib.cornell.edu/AgCensus/getVolumeTwoPart.do?volnum=5&year=1900&part_id=1096&number=1&title=Farms,%20Livestock,%20and%20Animal%20Products.

U.S. Census Bureau. "1860 Census: Agriculture of the United States." https://www.census.gov/library/publications/1864/dec/1860b.html.

———. "1860 Census: Population of the United States." https://www.census.gov/library/publications/1864/dec/1860a.html.

———. "1870 Census: A Compendium of the Ninth Census." https://www.census.gov/library/publications/1872/dec/1870e.html.

———. "Population of the 100 Largest Urban Places: 1860." https://www.census.gov/population/www/documentation/twps0027/tab09.txt.

U.S. Department of Agriculture. "Dietary Reference Intakes (DRI): Electrolytes and Water." http://nationalacademies.org/hmd/~/media/Files/Activity%20Files/Nutrition/DRI-Tables/9_Electrolytes_Water%20Summary.pdf?la=en. Accessed June 3, 2019.

———. *Hog Cholera: Its History, Nature, and Treatment, as Determined by the Inquiries and Investigations of the Bureau of Animal Industry*. Washington, DC: U.S. Government Printing Office, 1889.

U.S. Geological Survey and Tennessee Valley Authority, Mapping Services Branch. *Paducah East Quadrangle, Kentucky-Illinois: 7.5 Minute Series*. Reston, VA: U.S. Geological Survey, 1983.

———. *Shooks Gap Quadrangle, Tennessee: 7.5 Minute Series*. Reston, VA: U.S. Geological Survey, 1987.

U.S. War Department. *The War of the Rebellion: A Compilation of the Official Records of the Union and Confederate Armies*. 128 vols. Washington, DC: U.S. Government Printing Office, 1880–1901.

Van Buren, W. H. *Rules for Preserving the Health of the Soldier*. Washington, DC: Sanitary Commission, 1861.

Vance, Zebulon Baird. *The Papers of Zebulon Baird Vance*. Edited by Gordon B. McKinney and Richard M. McMurray. Microfilm. Frederick, MD: University Publications of America, 1987.

Voris, Alvin C. *A Citizen Soldier's Civil War: The Letters of Brevet Major General Alvin C. Voris*. Edited by Jerome Mushkat. Dekalb: Northern Illinois University Press, 2002.

Wainwright, Charles S. *A Diary of Battle: The Personal Journals of Colonel Charles S. Wainwright, 1861–1865*. Edited by Allan Nevins. New York: Harcourt, Brace & World, 1962.

Walker, Frances A., comp. *Reports on the Production of Agriculture in the Tenth Census (June 1, 1880)*. Washington, DC: U.S. Government Printing Office, 1883. http://www.agcensus.usda.gov/Publications/Historical_Publications/1880/1880a_v3-02.pdf.

———. *The Statistics of the Wealth and Industry of the United States, 1870*. Washington, DC: U.S. Government Printing Office, 1883.

Woodward, Joseph Javier. *Outlines of the Chief Camp Diseases of the United States Armies: As Observed during the Present War*. Philadelphia: Lippincott, 1863.

SECONDARY SOURCES

Books

Adams, Michael C. C. *Living Hell: The Dark Side of the Civil War*. Baltimore: Johns Hopkins University Press, 2015.

Armistead, Gene C. *Horses and Mules in the Civil War: A Complete History with a Roster of More than 700 War Horses*. Jefferson, NC: McFarland, 2013.

Ash, Stephen V. *When the Yankees Came: Conflict and Chaos in the Occupied South, 1861–1865*. Chapel Hill: University of North Carolina Press, 1995.

Atack, Jeremy, and Fred Bateman. *To Their Own Soil: Agriculture in the Antebellum North*. Henry A. Wallace Series on Agricultural History and Rural Studies. Ames: Iowa State University Press, 1987.

Ballard, Michael B. *The Civil War in Mississippi: Major Campaigns and Battles*. Oxford: University of Mississippi Press, 2011.

———. *Vicksburg: The Campaign That Opened the Mississippi*. Chapel Hill: University of North Carolina Press, 2004.

Barney, William L., ed. *The Oxford Encyclopedia of the Civil War*. New York: Oxford University Press, 2011.

Beirer, Bert W. *A Short History of Veterinary Medicine in America*. Lansing: Michigan State University Press, 1955.

Bell, Andrew McIlwaine. *Mosquito Soldiers: Malaria, Yellow Fever, and the Course of the American Civil War*. Baton Rouge: Louisiana State University Press, 2010.

Blair, William. *Virginia's Private War: Feeding Body and Soul in the Confederacy, 1861–1865*. New York: Oxford University Press, 1998.

Bollet, Alfred J. *Civil War Medicine: Challenges and Triumphs*. Tucson, AZ: Galen, 2002.

Bonekemper, Edward H. *A Victor, Not a Butcher: Ulysses S. Grant's Overlooked Military Genius*. Washington, DC: Regnery, 2004.

Borden, Skye. *Thirsty City: Politics, Greed, and the Making of Atlanta's Water Crisis*. Albany: State University of New York Press, 2014.

Brady, Lisa M. *War upon the Land: Military Strategy and the Transformation of Southern Landscapes during the American Civil War*. Athens: University of Georgia Press, 2012.

Bray, R. S. *Armies of Pestilence: The Impact of Disease on History*. Cambridge, UK: James Clarke, 2004.

Brown, Andrew. *Geology and the Gettysburg Campaign*. Commonwealth of Pennsylvania Department of Conservation and Natural Resources, Bureau of Topographic and Geologic Survey, Educational Series 5 (1962, revised 1997). http://www.docs.dcnr.pa.gov/cs/groups/public/documents/document/dcnr_014596.pdf.

Brown, D. Alexander. *Grierson's Raid: A Cavalry Adventure of the Civil War*. Urbana: University of Illinois Press, 1954.

Browning, Judkin. *The Seven Days' Battles, 1862: The War Begins Anew*. Santa Barbara, CA: Praeger, 2012.

———. *Shifting Loyalties: The Union Occupation of Eastern North Carolina*. Chapel Hill: University of North Carolina Press, 2011.
Burton, Brian K. *Extraordinary Circumstances: The Seven Days Battles*. Bloomington: Indiana University Press, 2001.
Carter, Samuel. *The Siege of Atlanta, 1864*. New York: St. Martin's, 1973.
Cashin, Joan E. *War Stuff: The Struggle for Human and Environmental Resources in the American Civil War*. New York: Cambridge University Press, 2018.
Castel, Albert. *Decision in the West: The Atlanta Campaign of 1864*. Lawrence: University Press of Kansas, 1992.
Cimprich, John. *Fort Pillow, a Civil War Massacre, and Public Memory*. Baton Rouge: Louisiana State University Press, 2005.
Clausewitz, Carl von. *On War*. Translated by Michael Howard and Peter Paret. New York: Oxford University Press, 1976.
Coco, Gregory. *A Strange and Blighted Land, Gettysburg: The Aftermath of a Battle*. Gettysburg, PA: Thomas, 1995.
Coggins, Allen R. *Tennessee Tragedies: Natural, Technological, and Societal Disasters in the Volunteer State*. Knoxville: University of Tennessee Press, 2011.
Collingham, Lizzie. *The Taste of War: World War II and the Battle for Food*. New York: Penguin, 2011.
Cooling, Benjamin Franklin. *Forts Henry and Donelson: The Key to the Confederate Heartland*. Knoxville: University of Tennessee Press, 1987.
Cozzens, Peter. *No Better Place to Die: The Battle of Stones River*. Urbana: University of Illinois Press, 1990.
———. *Shenandoah, 1862: Stonewall Jackson's Valley Campaign*. Chapel Hill: University of North Carolina Press, 2008.
Crawford, Martin. *Ashe County's Civil War: Community and Society in the Appalachian South*. Charlottesville: University of Virginia Press, 2001.
Cronon, William. *Nature's Metropolis: Chicago and the Great West*. New York: W. W. Norton, 1991.
Crosby, Alfred W. *Ecological Imperialism: The Biological Expansion of Europe, 900–1900*. Cambridge, UK: Cambridge University Press, 2004.
Cunningham, Horace H. *Doctors in Gray: The Confederate Medical Service*. Baton Rouge: Louisiana State University Press, 1993.
———. *Field Medical Services at the Battle of Manassas*. Athens: University of Georgia Press, 1968.
Davis, Paul K. *Besieged: 100 Great Sieges from Jericho to Sarajevo*. New York: Oxford University Press, 2001.
Dean, Adam Wesley. *An Agrarian Republic: Farming, Antislavery Politics, and Nature Parks in the Civil War Era*. Chapel Hill: University of North Carolina Press, 2016.
Devine, Shauna. *Learning from the Wounded: The Civil War and the Rise of American Medical Science*. Chapel Hill: University of North Carolina Press, 2014.
Dobson, Mary. *Murderous Contagion: A Human History of Disease*. London: Quercus, 2015.
Donahue, Brian. *The Great Meadow: Farmers and the Land in Colonial Concord*. New Haven, CT: Yale University Press, 2004.
Dougherty, Kevin. *The Peninsula Campaign: A Military Analysis*. Oxford: University Press of Mississippi, 2010.

Dowdey, Clifford. *The Seven Days: The Emergence of Lee*. Boston: Little, Brown, 1964.

Downs, Jim. *Sick from Freedom: African American Illness and Suffering during the Civil War and Reconstruction*. New York: Oxford University Press, 2012.

Drake, Brian Allen, ed. *The Blue, the Gray, and the Green: Toward an Environmental History of the Civil War*. Athens: University of Georgia Press, 2015.

Duncan, Dayton. *Seed of the Future: Yosemite and the Evolution of the National Park Idea*. Yosemite Conservancy, 2013.

Earley, Gerald L. *The Second United States Sharpshooters in the Civil War: A History and Roster*. Jefferson, NC: McFarland, 2009.

Eicher, David J. *The Longest Night: A Military History of the Civil War*. New York: Simon & Schuster, 2002.

Engle, Stephen D. *Yankee Dutchman: The Life of Franz Sigel*. Fayetteville: University of Arkansas Press, 1993.

Ernst, Kathleen. *Too Afraid to Cry: Maryland Civilians in the Antietam Campaign*. Mechanicsburg, PA: Stackpole, 1999.

Estabrook, Barry. *Pig Tales: An Omnivore's Quest for Sustainable Meat*. New York: W. W. Norton, 2015.

Fagan, Brian. *The Little Ice Age: How Climate Made History, 1300–1850*. New York: Basic Books, 2001.

Faust, Drew Gilpin. *This Republic of Suffering: Death and the American Civil War*. New York: Alfred A. Knopf, 2008.

Fellman, Michael. *Inside War: The Guerrilla Conflict in Missouri during the American Civil War*. New York: Oxford University Press, 1989.

Fenn, Elizabeth A. *Pox Americana: The Great Smallpox Epidemic of 1775–82*. New York: Hill and Wang, 2001.

Fiege, Mark. *The Republic of Nature: An Environmental History of the United States*. Seattle: University of Washington Press, 2012.

Foote, Shelby. *The Civil War: A Narrative*. 3 vols. New York: Random House, 1954–74.

Frank, Lisa Yvette Tendrich, ed. *The World of the Civil War: A Daily Life Encyclopedia*, vol. 1. Santa Barbara, CA: Greenwood, 2015.

Frazier, Donald S. *Blood and Treasure: Confederate Empire in the Southwest*. College Station: Texas A&M University Press, 1996.

Freemon, Frank R. *Gangrene and Glory: Medical Care during the American Civil War*. Madison, NJ: Fairleigh Dickinson University Press, 1998.

Furguson, Ernest B. *Chancellorsville, 1863: Souls of the Brave*. New York: Vintage, 1992.

Garrett, Franklin M. *Atlanta and Environs: A Chronicle of Its People and Events, 1880s–1930s*, vol. 2. Athens: University of Georgia Press, 1969.

Gates, Paul W. *Agriculture and the Civil War*. New York: Knopf, 1965.

———. *The Farmer's Age: Agriculture, 1815–1860*. Vol. 3 of *The Economic History of the United States*. White Plains, NY: M. E. Sharpe, 1960.

Gillet, Mary C. *The Army Medical Department, 1818–1865*. Washington, DC: Center of Military History, 1987.

Glatthaar, Joseph T. *Soldiering in the Army of Northern Virginia: A Statistical Portrait of the Troops Who Served under Robert E. Lee*. Chapel Hill: University of North Carolina Press, 2011.

Gott, Kendall D. *Where the South Lost the War: An Analysis of the Fort Henry–Fort Donelson Campaign, February 1862*. Mechanicsburg, PA: Stackpole, 2003.

Greene, Ann Norton. *Horses at Work: Harnessing Power in Industrial America*. Cambridge, MA: Harvard University Press, 2008.

Grimsley, Mark. *The Hard Hand of War: Union Military Policy toward Southern Civilians, 1861–1865*. New York: Cambridge University Press, 1995.

Groeling, Meg. *The Aftermath of Battle: The Burial of the Civil War Dead*. El Dorado Hills, CA: Savas Beatie, 2015.

Guelzo, Allen C. *Gettysburg: The Last Invasion*. New York: Alfred A. Knopf, 2013.

Hall, Charles W. L., ed. *Plowshares to Bayonets . . . In the Defense of the Heartland: A History of the 27th Regiment Mississippi Infantry, CSA*. Bloomington, IN: Trafford, 2012.

Handley-Cousins, Sarah. *Bodies in Blue: Disability in the Civil War North*. Athens: University of Georgia Press, 2019.

Hattaway, Herman. *Reflections of a Civil War Historian: Essays on Leadership, Society, and the Art of War*. Columbia: University of Missouri Press, 2004.

Henlein, Paul C. *Cattle Kingdom in the Ohio Valley, 1783–1860*. Lexington: University of Kentucky Press, 1959.

Hennessy, John J. *Return to Bull Run: The Campaign and Battle of Second Manassas*. New York: Simon & Schuster, 1993.

Hilliard, Sam Bowers. *Hog Meat and Hoecake: Food Supply in the Old South, 1840–1860*. Carbondale: Southern Illinois University Press, 1972. Reprint, Athens: University of Georgia Press, 2014.

Hoehling, A. A. *Vicksburg: 47 Days of Siege*. Englewood Cliffs, NJ: Prentice-Hall, 1969.

Humphreys, Margaret. *Intensely Human: The Health of the Black Soldier in the American Civil War*. Baltimore: Johns Hopkins University Press, 2008.

———. *Malaria: Poverty, Race, and Public Health in the United States*. Baltimore: Johns Hopkins University Press, 2001.

———. *Marrow of Tragedy: The Health Crisis of the American Civil War*. Baltimore: Johns Hopkins University Press, 2013.

———. *Yellow Fever and the South*. Baltimore: Johns Hopkins University Press, 1992.

Hurt, R. Douglas. *Agriculture and the Confederacy: Policy, Productivity, and Power in the Civil War South*. Chapel Hill: University of North Carolina Press, 2015.

Johnson, Gerald H. *Geology of the Yorktown, Poquoson West, and Poquoson East Quadrangles, Virginia*. Report of Investigations 30. Charlottesville: Virginia Division of Mineral Resources, 1972.

Johnson, Ludwell H. *Red River Campaign: Politics and Cotton in the Civil War*. Kent, OH: Kent State University Press, 1993.

Jones, Archer. *Civil War Command and Strategy: The Process of Victory and Defeat*. New York: Free Press, 1992.

Jordan, Brian Matthew. *Marching Home: Union Veterans and Their Unending Civil War*. New York: Liveright, 2014.

Josephy, Alvin M., Jr. *The Civil War in the American West*. New York: Alfred A. Knopf, 1991.

Kennett, Lee. *Marching through Georgia: The Story of Soldiers and Civilians during Sherman's Campaign*. New York: Harper Collins, 1995.

Kerr, Paul B. *Civil War Surgeon: Biography of James Langstaff Dunn, MD*. N.p.: Authorhouse, 2005.

Krick, Robert K. *Civil War Weather in Virginia*. Tuscaloosa: University of Alabama Press, 2007.

Leckie, William H., and Shirley A. Leckie. *Unlikely Warriors: General Benjamin Grierson and His Family*. Norman: University of Oklahoma Press, 1984.

Link, William A. *Atlanta, Cradle of the New South: Race and Remembering in the Civil War's Aftermath*. Chapel Hill: University of North Carolina Press, 2013.

Lonn, Ella. *Desertion during the Civil War*. New York: Century, 1928.

———. *Salt as a Factor in the Confederacy*. New York: W. Neale, 1933.

Lowry, Thomas P. *The Story the Soldiers Wouldn't Tell: Sex in the Civil War*. Mechanicsburg, PA: Stackpole, 1994.

Manarin, Louis H., and Weymouth T. Jordan, comps. *North Carolina Troops: A Roster*. 18 vols. Raleigh: North Carolina Division of Archives and History, 1966–.

Marten, James Alan. *Sing Not War: The Lives of Union & Confederate Veterans in Gilded Age America*. Chapel Hill: University of North Carolina Press, 2011.

Marvel, William. *Burnside*. Chapel Hill: University of North Carolina Press, 1991.

———. *A Place Called Appomattox*. Chapel Hill: University of North Carolina Press, 2000.

Masich, Andrew Edward. *The Civil War in Arizona: The Story of the California Volunteers, 1861–1865*. Norman: University of Oklahoma Press, 2006.

Massey, Mary Elizabeth. *Ersatz in the Confederacy*. Columbia, SC: University of South Carolina Press, 1952.

Matter, William D. *If It Takes All Summer: The Battle of Spotsylvania*. Chapel Hill: University of North Carolina Press, 1988.

Mauldin, Erin Stewart. *Unredeemed Land: An Environmental History of Civil War and Emancipation in the Cotton South*. New York: Oxford University Press, 2018.

McCurry, Stephanie. *Confederate Reckoning: Power and Politics in the Civil War South*. Cambridge, MA: Harvard University Press, 2010.

McDonough, James Lee. *Chattanooga: A Death Grip on the Confederacy*. Knoxville: University of Tennessee Press, 1984.

———. *Shiloh: In Hell before Night*. Knoxville: University of Tennessee Press, 1977.

McGaugh, Scott. *Surgeon in Blue: Jonathan Letterman, the Civil War Doctor Who Pioneered Battlefield Care*. New York: Arcade, 2013.

McNeill, J. R. *Mosquito Empires: Ecology and War in the Greater Caribbean, 1620–1914*. Cambridge, UK: Cambridge University Press, 2010.

McPherson, James M. *Battle Cry of Freedom: The Civil War Era*. New York: Oxford University Press, 1988.

———. *Crossroads of Freedom: Antietam*. New York: Oxford University Press, 2002.

McPherson, James M., and James K. Hogue. *Ordeal by Fire: The Civil War and Reconstruction*. 4th ed. Boston: McGraw-Hill, 2009.

McShane, Clay, and Joel A. Tarr. *The Horse in the City: Living Machines of the Nineteenth Century*. Baltimore: Johns Hopkins University Press, 2007.

Meier, Kathryn Shively. *Nature's Civil War: Common Soldiers and the Environment in 1862 Virginia*. Chapel Hill: University of North Carolina Press, 2013.

Moore, Jerrod Northrup. *Confederate Commissary General: Lucius Bellinger Northrop and the Subsistence Bureau of the Southern Army*. Shippensburg, PA: White Mane, 1996.

Morris, Christopher. *Becoming Southern: The Evolution of a Way of Life, Warren County and Vicksburg, Mississippi, 1770–1860*. New York: Oxford University Press, 1995.

Morris, Roy, Jr. *The Better Angel: Walt Whitman in the Civil War*. New York: Oxford University Press, 2001.

Murray, Williamson, and Wayne Wei-siang Hsieh. *A Savage War: A Military History of the Civil War*. Princeton, NJ: Princeton University Press, 2016.

National Park Service, Geologic Resources Division. *Stones River National Battlefield: Geologic Resources Report*. Natural Resource Report NPS/NRSS/GRD/NRR—2012/566, 2012. https://irma.nps.gov/DataStore/DownloadFile/454639.

Nelson, Megan Kate. *Ruin Nation: Destruction and the American Civil War*. Athens: University of Georgia Press, 2012.

Nelson, Scott Reynolds, and Carol Sheriff. *A People at War: Civilians and Soldiers in America's Civil War, 1854–1877*. New York: Oxford University Press, 2008.

Noe, Kenneth W. *Perryville: This Grand Havoc of Battle*. Lexington: University Press of Kentucky, 2001.

Obreiter, John. *The Seventy-Seventh Pennsylvania at Shiloh: History of the Regiment*. Harrisburg, PA: Harrisburg, 1905.

Otto, John Solomon. *Southern Agriculture during the Civil War Era, 1860–1880*. Westport, CT: Greenwood, 1994.

Pearson, Chris. *Scarred Landscapes: War and Nature in Vichy France*. New York: Palgrave Macmillan, 2008.

Pfanz, Harry W. *Gettysburg: Culp's Hill and Cemetery Hill*. Chapel Hill: University of North Carolina Press, 1993.

Piston, William Garrett, and Richard W. Hatcher III. *Wilson's Creek: The Second Battle of the Civil War and the Men Who Fought It*. Chapel Hill: University of North Carolina Press, 2000.

Plumb, Robert C. *Your Brother in Arms: A Union Soldier's Odyssey*. Columbia: University of Missouri Press, 2013.

Poe, Clarence, ed. *True Tales of the South at War: How Soldiers Fought and Families Lived, 1861–1865*. Chapel Hill: University of North Carolina Press, 1961.

Ponting, Clive. *The Crimean War: The Truth behind the Myth*. London: Chatto & Windus, 2004.

Powell, David A. *The Chickamauga Campaign*. El Dorado Hills, CA: Savas Beatie, 2014.

Priest, John Michael. *Antietam: The Soldier's Battle*. Shippensburg, PA: White Mane, 1989.

———. *Before Antietam: The Battle for South Mountain*. New York: Oxford University Press, 1996.

Rable, George C. *Fredericksburg! Fredericksburg!* Chapel Hill: University of North Carolina Press, 2003.

———. *God's Almost Chosen People: A Religious History of the American Civil War*. Chapel Hill: University of North Carolina Press, 2010.

Raines, Rebecca Robbins. *Getting the Message Through: A Branch History of the Army Signal Corps*. Army Historical Series. Center of Military History, United States Army. Washington, DC: U.S. Government Printing Office, 1999.

Rhea, Gordon C. *The Battle of the Wilderness, May 5–6, 1864*. Baton Rouge: Louisiana State University Press, 1994.

———. *The Battles for Spotsylvania Court House and the Road to Yellow Tavern, May 7–12, 1864*. Baton Rouge: Louisiana State University Press, 1997.

———. *Cold Harbor: Grant and Lee, May 26–June 3, 1864*. Baton Rouge: Louisiana State University Press, 2002.

Rosenberg, Charles E. *The Cholera Years: The United States in 1832, 1849, and 1866*. 2nd ed. Chicago: University of Chicago Press, 1987.

Ruhlman, R. Fred. *Captain Henry Wirz and Andersonville Prison: A Reappraisal*. Knoxville: University of Tennessee Press, 2006.
Runte, Alfred. *National Parks: The American Experience*. Lincoln: University of Nebraska Press, 1979.
Salling, Stuart. *Louisianans in the Western Confederacy: The Adams-Gibson Brigade in the Civil War*. Jefferson, NC: McFarland, 2010.
Schneider, Paul. *Old Man River: The Mississippi River in North American History*. New York: Henry Holt, 2013.
Schroeder-Lein, Glenna R. *The Encyclopedia of Civil War Medicine*. London: Routledge, 2015.
Sears, Stephen W. *Chancellorsville*. Boston: Mariner, 1996.
———. *George B. McClellan: The Young Napoleon*. New York: Da Capo, 1999.
———. *Gettysburg*. Boston: Houghton-Mifflin, 2003.
———. *Landscape Turned Red: The Battle of Antietam*. Boston: Mariner, 1983.
———. *To the Gates of Richmond: The Peninsula Campaign*. New York: Ticknor & Fields, 1992.
Sharrer, G. Terry. *A Kind of Fate: Agricultural Change in Virginia, 1861–1920*. Lafayette, IN: Purdue University Press, 2002.
Sherwood, Lauralee. *Human Physiology: From Cells to Systems*. 8th ed. Belmont, CA: Brooks/Cole, Cengage Learning, 2013.
Silver, Timothy. *Mount Mitchell and the Black Mountains: An Environmental History of the Highest Peaks in Eastern America*. Chapel Hill: University of North Carolina Press, 2003.
———. *A New Face on the Countryside: Indians, Colonists, and Slaves in South Atlantic Forests, 1500–1800*. New York: Cambridge University Press, 1990.
Smith, Andrew F. *Starving the South: How the North Won the Civil War*. New York: St. Martin's, 2011.
Smith, Mark M. *The Smell of Battle, the Taste of Siege: A Sensory History of the Civil War*. New York: Oxford University Press, 2015.
Smith, Timothy B. *The Untold Story of Shiloh: The Battle and the Battlefield*. Knoxville: University of Tennessee Press, 2006.
Soper, Steve. *The "Glorious Old Third": A History of the Third Michigan Infantry, 1855 to 1927*. [Michigan]: Old Third, 1998–2007. www.books.google.com/books?isbn=0978786106.
Starr, Stephen Z. *The Union Cavalry in the Civil War*. Vol. 2, *The War in the East: From Gettysburg to Appomattox, 1863–1865*. Baton Rouge: Louisiana State University Press, 1981.
Steinberg, Ted. *Down to Earth: Nature's Role in American History*. New York: Oxford University Press, 2002.
Steiner, Paul. *Disease and the Civil War: Natural Biological Warfare in 1861–1865*. Springfield, IL: Charles C. Thomas, 1968.
Sternhell, Yael A. *Routes of War: The World of Movement in the Confederate South*. Cambridge, MA: Harvard University Press, 2012.
Stewart, Mart A. *What Nature Suffers to Groe: Life, Labor, and Landscape on the Georgia Coast, 1680–1920*. Athens: University of Georgia Press, 2002.
Stith, Matthew M. *Extreme Civil War: Guerrilla Warfare, Environment, and Race on the Trans-Mississippi Frontier*. Baton Rouge: Louisiana State University Press, 2016.
Stoll, Steven. *Larding the Lean Earth: Soil and Society in Nineteenth-Century America*. New York: Hill and Wang, 2002.

Strom, Claire. *Making Catfish Bait Out of Government Boys: The Fight against Cattle Ticks and the Transformation of the Yeoman South*. Athens: University of Georgia Press, 2009.

Sutherland, Daniel E. *A Savage Conflict: The Decisive Role of Guerrillas in the American Civil War*. Chapel Hill: University of North Carolina Press, 2009.

———. *Seasons of War: The Ordeal of a Confederate Community, 1861–1865*. Baton Rouge: Louisiana State University Press, 1995.

Sutter, Paul S. *Let Us Now Praise Famous Gullies: Providence Canyon and the Soils of the South*. Athens: University of Georgia Press, 2015.

Taylor, Amy Murrell. *Embattled Freedom: Journeys through the Civil War's Slave Refugee Camps*. Chapel Hill: University of North Carolina Press, 2018.

Thompson, Jerry D. *Confederate General of the West: Henry Hopkins Sibley*. Natchitoches, LA: Northwestern State University Press, 1987. Reprint, College Station: Texas A&M University Press, 1996.

Todd, Richard Cecil. *Confederate Finance*. Athens: University of Georgia Press, 1954.

Trudeau, Noah Andre. *Bloody Roads South: The Wilderness to Cold Harbor, May–June, 1864*. Boston: Little, Brown, 1989.

———. *The Last Citadel: Petersburg, Virginia, June 1864–April 1865*. Boston: Little, Brown, 1991.

Turkington, Carol, and Bonnie Lee Ashby, M.D., eds. *The Encyclopedia of Infectious Diseases*. 3rd ed. New York: Facts on File, 2007.

Varon, Elizabeth R. *Appomattox: Victory, Defeat, and Freedom at the End of the Civil War*. New York: Oxford University Press, 2014.

Weigley, Russell. *A Great Civil War: A Military and Political History*. Bloomington: Indiana University Press, 2000.

Wert, Jeffrey D. *General James Longstreet, the Confederacy's Most Controversial Soldier: A Biography*. New York: Simon & Schuster, 1993.

Wheeler, Richard. *The Siege of Vicksburg*. New York: Thomas Y. Crowell, 1978.

Whisonant, Robert C. *Arming the Confederacy: How Virginia's Minerals Forged the Rebel War Machine*. Cham, Switzerland: Springer International, 2015.

Wiley, Bell Irvin. *The Life of Johnny Reb: The Common Soldier of the Confederacy*. Indianapolis: Bobbs-Merrill, 1943. Reprint, Baton Rouge: Louisiana State University Press, 2008.

Williams, David. *A People's History of the Civil War: Struggles for the Meaning of Freedom*. New York: New Press, 2005.

Williams, T. Harry. *Lincoln and His Generals*. New York: Knopf, 1952.

Wills, John. *Conservation Fallout: Nuclear Protest at Diablo Canyon*. Reno: University of Nevada Press, 2006.

Wilson, Mark R. *The Business of Civil War: Military Mobilization and the State, 1861–1865*. Baltimore: Johns Hopkins University Press, 2006.

Winters, Harold A., with Gerald E. Galloway Jr., William J. Reynolds, and David W. Rhye. *Battling the Elements: Weather and Terrain in the Conduct of War*. Baltimore: Johns Hopkins University Press, 1998.

Wise, Jennings C. *The Long Arm of Lee: The History of the Artillery of the Army of Northern Virginia*. Lynchburg, VA: J. P. Bell, 1915. Reprint, New York: Oxford University Press, 1959.

Woodworth, Steven E. *Nothing but Victory: The Army of the Tennessee, 1861–1865*. New York: Alfred A. Knopf, 2005.

Woolridge, William C. *Mapping Virginia: From the Age of Exploration to the Civil War*. Charlottesville: University of Virginia Press, 2012.

Wortman, Marc. *The Bonfire: The Siege and Burning of Atlanta*. New York: PublicAffairs, 2009.

Young, Alfred C., III. *Lee's Army during the Overland Campaign: A Numerical Study*. Baton Rouge: Louisiana State University Press, 2013.

Articles and Chapters

"2011 Compendium of Physical Activity." https://sites.google.com/site/compendium ofphysicalactivities/Activity-Categories/walking.

Abbot, Karen. "A Newspaper for Injured Civil War Vets." *New York Times*, Disunion blog, December 22, 2014. http://opinionator.blogs.nytimes.com/2014/12/22/a-newspaper-for-injured-civil-war-vets/.

Adams, Sean Patrick. "Iron from the Wilderness: The History of Virginia's Catharine Furnace." Fredericksburg and Spotsylvania National Military Park Historic Resource Study. National Park Service, U.S. Department of the Interior, Northeast Region History Program, June 2011.

Alexander, Ted. "Destruction, Disease, and Death: The Battle of Antietam and the Sharpsburg Civilians." *Civil War Regiments* 6, no. 2 (1998): 148–55.

Allison, Roger, Jim Glanville, and Harry Haynes. "Saltville during the Civil War." *Journal of the Historical Society of Western Virginia* 20 (2012): 73–82.

Altonen, Brian. "1851–1917, Cattle Drives and Texas Fever." http://brianaltonenmph.com/gis/historical-disease-maps/zoonoses/1866-1885-the-texas-cattle-drives-and-texas-fever. Accessed May 2, 2015.

Alvarez-Ordonez, Avelino, Maire Begley, Miguel Prieto, Winy Messens, Mercedes Lopez, Ana Bernardo, and Colin Hill. "Salmonella spp. Survival Strategies within the Host Gastrointestinal Tract." *Microbiology* 157 (2011): 3268–81.

Aryal, Sagar. "Differences between Diarrhea and Dysentery." June 11, 2015. http://www.microbiologyinfo.com/differences-between-diarrhea-and-dysentery/.

Bass, S. Jonathan. "'How 'bout a Hand for the Hog': The Enduring Nature of the Swine as a Cultural Symbol in the South." *Southern Cultures* 1, no. 3 (Spring 1995): 301–20.

Bell, Andrew McIlwaine. "'Gallinippers' & Glory: The Links between Mosquito-Borne Disease and U.S. Civil War Operations and Strategy, 1862." *Journal of Military History* 74 (April 2010): 379–405.

Berry, Stephen. "Forum: The Future of Civil War Studies." *Journal of the Civil War Era* 2, no. 1 (March 2012). https://www.journalofthecivilwarera.org/forum-the-future-of-civil-war-era-studies/.

Black, Brian. "The Copse at Gettysburg." *Environmental History* 9, no. 2 (April 2004): 306–10.

Blevins, Brooks. "Cattle Raising in Antebellum Alabama." *Alabama Review* 51, no. 4 (October 1998): 270–88.

Bohannon, Keith. "'Dirty, Ragged, and Ill-Provided For': Confederate Logistical Problems in the 1862 Maryland Campaign and Their Solutions." In *The Antietam Campaign*, edited by Gary W. Gallagher, 101–42. Chapel Hill: University of North Carolina Press, 1999.

Bonhotal, Jean, Mary Schwarz, Craig Williams, and Ann Swinker. "Horse Mortality:

    Carcass Disposal Alternatives." Cornell Waste Management Institute, 2012. http://cwmi.css.cornell.edu/horsefs.pdf.

Brady, Lisa M. "Nature as Friction: Integrating Clausewitz into Environmental Histories of the Civil War." In *The Blue, the Gray, and the Green: Toward an Environmental History of the Civil War*, edited by Brian Allen Drake, 144–62. Athens: University of Georgia Press, 2015.

———. "The Wilderness of War: Nature and Strategy in the American Civil War." In *Environmental History and the American South: A Reader*, edited by Paul S. Sutter and Christopher Manganiello, 168–95. Athens: University of Georgia Press, 2009.

Brady, Lisa M., and Timothy Silver. "Nature as Material Culture: Antietam National Battlefield." In *War Matters: Material Culture in the Civil War Era*, edited by Joan E. Cashin, 53–74. Chapel Hill: University of North Carolina Press, 2018.

Browning, Judkin. "All for One Charge: The 44th Georgia Infantry." *Columbiad* 1, no. 4 (Winter 1998): 21–45.

Burt, Christopher C. "California's Superstorm: The USGS ARkStorm Report and the Great Flood of 1862." https://www.wunderground.com/blog/weatherhistorian/californias-superstorm-the-usgs-arkstorm-report-and-the-great-flood-.html. Accessed October 6, 2017.

Byrne, James. "Divine Intervention via a Microbe." Guest blog, *Scientific American*, November 19, 2010. https://blogs.scientificamerican.com/guest-blog/divine-intervention-via-a-microbe/.

"Calomel." *Encyclopaedia Britannica*. https://www.britannica.com/science/calomel. Accessed October 12, 2015.

Centers for Disease Control. "Measles." In *Epidemiology and Prevention of Vaccine-Preventable Diseases: The Pink Book, Course Textbook*, 13th ed. (April 2015): 209–14. http://www.cdc.gov/vaccines/pubs/pinkbook/downloads/meas.pdf.

Centers for Disease Control and Prevention. "Human Factors and Malaria." https://www.cdc.gov/malaria/about/biology/human_factors.html. Accessed October 20, 2015.

———. "Measles (Rubeola)." http://www.cdc.gov/measles/hcp/. Accessed October 15, 2015.

———. "Question and Answers: What Is Shigellosis?" http://www.cdc.gov/shigella/general-information.html#transmission. Accessed October 10, 2015.

"Chaos." *Stanford Encyclopedia of Philosophy*. http://plato.stanford.edu/entries/chaos/#DefChaDetNonSenDep. Accessed November 8, 2014.

Chernin, Eli. "Surgical Maggots." *Southern Medical Journal* 79, no. 9 (September 1986): 1143–45.

"Civil War Casualties." http://www.civilwar.org/education/civil-war-casualties.html?referrer=https://www.google.com/. Accessed February 12, 2017.

"Civil War Medical Terms." *ehistory*, Department of History, The Ohio State University. https://ehistory.osu.edu/exhibitions/cwsurgeon/cwsurgeon/medicalterms. Accessed October 1, 2017.

Clinton, Catherine. "'Public Women' and Sexual Politics during the American Civil War." In *Battle Scars: Gender and Sexuality in the American Civil War*, edited by Catherine Clinton and Nina Silber, 61–77. New York: Oxford University Press, 2006.

Cohen, Jennie. "Did Jamestown Settlers Drink Themselves to Death?" *History*, October

17, 2011. http://www.history.com/news/did-jamestowns-settlers-drink-themselves-to-death.

Conlon, Joseph M. "The Historical Impact of Epidemic Typhus." http://www.montana.edu/historybug/documents/TYPHUS-Conlon.pdf. Accessed October 2, 2015.

Costandi, Moheb. "Life after Death: The Science of Decomposition." *Guardian*, May 5, 2015. https://www.theguardian.com/science/neurophilosophy/2015/may/05/life-after-death.

———. "The Smell of Death: Scientist Reveals How 400 Compounds Mix to Create Heady Mixture of Scents as Bacteria Rips Apart Rotting Flesh." *Daily Mail* (UK), May 6, 2015. http://www.dailymail.co.uk/sciencetech/article-3071037/The-smell-death-Scientist-reveals-bacteria-creates-heady-mixture-scents-rips-apart-rotting-flesh.html.

Coulter, Nate. "The Impact of the Civil War upon Pulaski County, Arkansas." *Arkansas Historical Quarterly* 41, no. 1 (Spring 1982): 67–82.

Courtwright, David T. "Opiate Addiction as a Consequence of the Civil War." *Civil War History* 24, no. 2 (June 1978): 101–11.

Davis, Robert Scott. "Andersonville Prison." *New Georgia Encyclopedia*. http://www.georgiaencyclopedia.org/articles/history-archaeology/andersonville-prison. Accessed December 15, 2016.

Davis, Stephen. "'A Very Barbarous Mode of Carrying on War': Sherman's Artillery Bombardment of Atlanta, July 20–August 24, 1864." *Georgia Historical Quarterly* 79, no. 1 (Spring 1995): 57–90.

"Decomposition." Forensics Library. http://aboutforensics.co.uk/decomposition/. Accessed July 18, 2016.

DeGruccio, Michael. "Letting the War Slip through Our Hands: Material Culture and the Weakness of Words in the Civil War Era." In *Weirding the War: Stories from the Civil War's Ragged Edges*, edited by Stephen Berry, 15–35. Athens: University of Georgia Press, 2011.

Diamant, Rolf. "Lincoln, Olmsted, and Yosemite: Time for a Closer Look." *George Wright Forum* 31, no. 1 (2014): 10–16.

Dodman, Thomas. "1814 and the Melancholy of War." *Journal of Military History* 80, no. 1 (January 2016): 31–55.

Drake, Brian. "New Fields of Battle: Nature, Environmental History, and the Civil War." In *The Blue, the Gray, and the Green: Toward an Environmental History of the Civil War*, edited by Brian Allen Drake, 1–15. Athens: University of Georgia Press, 2015.

Earle, Carville. "Environment, Disease, and Mortality in Early Virginia." *Journal of Historical Geography* 5, no. 4 (February 1979): 365–90.

Ehlen, Judy, and R. C. Whisonant. "Military Geology of Antietam Battlefield, Maryland, USA—Geology, Terrain, and Casualties." *Geology Today* (January 30, 2008). http://onlinelibrary.wiley.com/doi/10.1111/j.1365-2451.2008.00647.x/full.

Elmore, Bartow. "Hydrology and Residential Segregation in the Postwar South: An Environmental History of Atlanta, 1865–1895." *Georgia Historical Quarterly* 94, no. 1 (Spring 2010): 30–61.

"El Niño & La Niña." http://www.weatherexplained.com/Vol-1/El-Ni-o-La-Ni-a.html. Accessed October 21, 2014.

Fales, Daniel C. "M-16: The Gun They Swear by . . . and At!" *Popular Mechanics* 128, no. 4 (October 1967): 128–31.

"Famous Quotes Concerning the National Parks." https://www.nps.gov/parkhistory/hisnps/npsthinking/famousquotes.htm. Accessed November 30, 2018.

Fiege, Mark. "Gettysburg and the Organic Nature of the American Civil War." In *Natural Enemy, Natural Ally: Toward an Environmental History of War*, edited by Richard Tucker and Edmund Russell, 93–109. Corvallis: Oregon State University Press, 2004.

Figg, Laurann, and Jane Farrell-Beck. "Amputation in the Civil War: Physical and Social Dimensions." *Journal of the History of Medicine and Allied Sciences* 48, no. 4 (October 1993): 454–75.

"Food Habits of Feral Hogs." *Extension*, October 9, 2012. http://www.extension.org/pages/63655/food-habits-of-feral-hogs#.VYhOVFVVikp.

Gagnon, Amy. "Death and Mourning in the Civil War Era." http://connecticuthistory.org/death-and-mourning-in-the-civil-war-era/. Accessed November 10, 2016.

Gardner, Douglas G. "Prisoners of War." In *Encyclopedia of the American Civil War: A Political, Social, and Military History*. 5 vols. Edited by David Heidler and Jeanne Heidler, 3:1572. Santa Barbara, CA: ABC-CLIO, 2000.

Gettysburg National Military Park. "History and Culture." https://www.nps.gov/gett/learn/historyculture/index.htm. Accessed November 30, 2018.

Graebner, Norman A. "Northern Diplomacy and European Neutrality." In *Why the North Won the Civil War*, edited by David Herbert Donald, 58–80. Baton Rouge: Louisiana State University Press, 1960. Reprint, New York: Simon & Schuster, 1996.

Grav, Hans-Petter. "When the Beast Saved the Day and Yellow Jack Got Lost: The Story of General Butler and the Yellow Fever Epidemic That Never Took Place." *Southern Historian* 33 (Spring 2012): 37–51.

Gray, Michael P. "Elmira, a City on a Prison-Camp Contract." *Civil War History* 44 (December 1999): 322–38.

Griffith, Ashley. "Grains for Horses and Their Characteristics." *Extension*, July 1, 2013. http://www.extension.org/pages/10246/grains-for-horses-and-their-characteristics#.VXYBOaby07A.

Guan, Bin, Duane E. Waliser, Noah P. Molotch, Eric J. Fetzer, and Paul J. Neiman. "Does the Madden-Julian Oscillation Influence Wintertime Atmospheric Rivers and Snowpack in the Sierra Nevada?" *Monthly Weather Review* 140, no. 12 (February 2012): 325–42.

Guinn, J. M. "Exceptional Years: A History of California's Floods & Drought." *Historical Society of Southern California, Los Angeles (1890)* 1, no. 5 (1890): 33–39.

Hacker, J. David. "A Census-Based Count of the Civil War Dead." *Civil War History* 57, no. 4 (December 2011): 307–48.

Hacker, J. David, Libra Hilde, and James Holland Jones. "The Effect of the Civil War on Southern Marriage Patterns." *Journal of Southern History* 76, no. 1 (February 2010): 39–70.

Hall, John B., and Susan Silver. "Nutrition and Feeding of the Calf-Cow Herd: Digestive System of the Cow." Virginia Cooperative Extension Publication 400-010 (2009). https://pubs.ext.vt.edu/400/400-010/400-010_pdf.pdf.

Hamidullah, Intisar K. "The Impact of Disease on the Civil War." Yale National Initiative to Strengthen Teaching in Public Schools. http://teachers.yale.edu/curriculum/viewer/initiative_10.06.02_u. Accessed October 27, 2015.

Hart, Gavin. "Sexual Behavior in a War Environment." *Journal of Sex Research* 11, no. 3 (August 1975): 218–20.

Haygood, Tamara Miner. "Cows, Ticks, and Disease: A Medical Interpretation of the Southern Cattle Industry." *Journal of Southern History* 52, no. 4 (November 1986): 551–64.

Heiser, John. "The Widow and Her Farm." Blog of Gettysburg National Military Park, January 8, 2015. https://npsgnmp.wordpress.com/2015/01/08/the-widow-and-her-farm/.

Herweijer, Celine, Richard Seager, and Edward R. Cook. "North American Droughts of the Mid- to Late-Nineteenth Century: A History Simulation and Implication for Mediaeval Drought." *Holocene* 16 (February 2006): 159–71.

Hong, Sok Chul. "The Burden of Early Exposure to Malaria in the United States, 1850–1860: Malnutrition and Immune Disorders." *Journal of Economic History* 67 (December 2007): 1001–35.

"Humid Continental Climate." *Encyclopedia Britannica.* http://www.britannnica.com/EBchecked/topic/276210/humid-continental-climate. Accessed September 10, 2014.

Humphreys, Margaret. "A Stranger to Our Camps: Typhus in American History." *Bulletin of the History of Medicine* 80 (Summer 2006): 270–71.

———. "This Place of Death: Environment as a Weapon in the American Civil War." *Southern Quarterly* 53 (Spring–Summer 2016): 12–36.

Hupy, Joseph P. "The Environmental Footprint of War." *Environment and History* 14, no. 3 (August 2008): 405–21.

Indiana Division of Forestry. "Livestock Grazing in Woodlands." http://www.in.gov/dnr/forestry/files/livestockgrazinginwoodlands.pdf. Accessed May 1, 2015.

Ingram, B. Lynn. "California Megaflood: Lessons from a Forgotten Catastrophe." *Scientific American* 308 (December 18, 2012). http://www.scientificamerican.com/article/atmospheric-rivers-california-megaflood-lessons-from-forgotten-catastrophe/.

"John Forney Zacharias." *Journal of the American Medical Association* 43, no. 11 (September 10, 1904): 748.

Jones, Jonathan S. "Opium Slavery: Civil War Veterans and Opiate Addiction." *Journal of the Civil War Era*, forthcoming 2020.

Jones, Spencer. "The Influence of Horse Supply upon Field Artillery in the American Civil War." *Journal of Military History* 74 (April 2010): 357–77.

Joseph, Isaac, Deepu G. Mathew, Pradeesh Sathyan, and Geetha Vargheese. "The Use of Insects in Forensic Investigations: An Overview on the Scope of Forensic Entomology." *Journal of Forensic Dental Sciences* 3, no. 2 (July 2011): 89–91. http://www.ncbi.nlm.nih.gov/pmc/articles/PMC3296382/.

Kemmerly, Phillip R. "Environment and the Course of Battle: Flooding at Shiloh (6–7 April, 1862)." *Journal of Military History* 79 (October 2015): 1079–108.

King, Janet. "Maggots: Friend or Foe?" Civil War Rx. http://civilwarrx.blogspot.com/2014/02/maggots-friend-or-foe.html. Accessed February 1, 2017.

Kirby, Jack Temple. "The American Civil War: An Environmental View." National Humanities Center. *Nature Transformed: The Environment in American History.* Revised July 2001. http://www.nhc.rtp.nc.us/tserve/nattrans/ntuseland/essays/amcwar.htm.

Kneipp, Gregg. "The Wilderness as Wilderness, Then and Now." In *Hell Itself: The Battle of the Wilderness, May 5–7, 1864*, edited by Chris Mackowski, 2320–28. Kindle ed. El Dorado Hills, CA: Savas-Beatie, 2016.

Krick, Robert K. "An Insurmountable Barrier between the Army and Ruin: The Confederate Experience at Spotsylvania's Bloody Angle." In *The Spotsylvania Campaign*, edited by Gary W. Gallagher, 80–126. Chapel Hill: University of North Carolina Press, 1998.

———. "'Lee to the Rear,' the Texans Cried." In *The Wilderness Campaign*, edited by Gary W. Gallagher, 160–200. Chapel Hill: University of North Carolina Press, 2006.

Kruglikova, A. A., and S. I. Chernysh. "Surgical Maggots and the History of Their Medical Use." *Entomological Review* 93, no. 6 (September 2013): 667–74.

"La Niña Fact Sheet: Feature Articles." http://earthobservatory.nasa.gov/Features/La Nina/la_nina_2.php. Accessed October 10, 2014.

Larson, C. Kay. "The Horses of War." *New York Times*, Opinionator blog, February 2, 2013. http://opinionator.blogs.nytimes.com/2013/02/02/the-horses-of-war/?_r=0.

Lee, James C. "The Undertaker's Role during the American Civil War." HistoryNet, June 12, 2006. http://www.historynet.com/the-undertakers-role-during-the-american-civil-war.htm.

Leigh, Phil. "Who Burned Atlanta?" *New York Times*, Civil War Disunion blog, November 13, 2014. https://opinionator.blogs.nytimes.com/2014/11/13/who-burned-atlanta/?_r=0.

Levin, Kevin M. "'The Devil Himself Could Not Have Checked Them': Fighting with Black Soldiers at the Crater." In *Cold Harbor to the Crater: The End of the Overland Campaign*, edited by Gary W. Gallagher and Caroline E. Janney, 264–82. Chapel Hill: University of North Carolina Press, 2015.

Lipman, Don. "Bull Run Battle 1861: As Hot Then as Now?" *Washington Post*, July 21, 2011. https://www.washingtonpost.com/blogs/capital-weather-gang/post/bull-run-battle-as-hot-then-as-now/2011/07/21/gIQAYDNtRI_blog.html?utm_term=.eb66d557f202#pagebreak.

"Little Ice Age: Environmental History Timeline." http://www.eh-resources.org/timeline/timeline_lia.html. Accessed October 10, 2014.

Mackowiak, Philip A., and Paul H. Sehdev. "The Origin of Quarantine." *Clinical Infectious Diseases* 35 (November 1, 2002): 1071–72. https://academic.oup.com/cid/article/35/9/1071/330421.

Marshall, Nicholas. "The Great Exaggeration: Death and the Civil War." *Journal of the Civil War Era* 4, no. 1 (March 2014): 3–27.

Maryland Cooperative Extension. "Ask the Experts." http://www.equiery.com/archives/AskTheExperts/8_7Experts_Bury.html. Accessed July 9, 2015.

Mauldin, Erin Stewart. "Freedom, Economic Autonomy, and Ecological Change in the Cotton South, 1865–1880." *Journal of the Civil War Era* 7 (September 2017): 401–24.

Mayo Clinic Staff. "Diseases and Conditions: Typhoid Fever, Causes." http://www.mayoclinic.org/diseases-conditions/typhoid-fever/basics/causes/con-20028553. Accessed September 20, 2015.

———. "E. Coli." http://www.mayoclinic.org/diseases-conditions/e-coli/basics/definition/con-20032105. Accessed October 10, 2015.

———. "Gonorrhea." http://www.nmihi.com/f/gonorrhea.htm. Accessed October 20, 2015.

———. "Salmonella Infection." http://www.mayoclinic.org/diseases-conditions/salmonella/basics/causes/con-20029017. Accessed October 10, 2015.

———. "Syphilis." http://www.mayoclinic.org/diseases-conditions/syphilis/symptoms-causes/dxc-20234443. Accessed October 20, 2015.

McMillen, Matt. "Beef." In *The New Encyclopedia of Southern Culture*, vol. 7. Edited by John T. Edge, 26. Chapel Hill: University of North Carolina Press, 2007.

Metcalf, John. "Inside the Patent Office's Cabinet of Ghoulish Inventions." *Atlantic*, October 26, 2015. http://www.citylab.com/design/2015/10/inside-the-patent-offices-cabinet-of-ghoulish-inventions/412276/.

Meyer, Nathan R. "Elmira Prison." In *Encyclopedia of the American Civil War: A Political, Social, and Military History*. 5 vols., edited by David Heidler and Jeanne Heidler, 2:648–49. Santa Barbara, CA: ABC-CLIO, 2000.

Miller, Gary L. "Historical Natural History: Insects and the Civil War." http://www.montana.edu/historybug/civilwar2/lousy.html. Accessed September 14, 2014.

Miller, William J. "I Only Wait for the River: McClellan and His Engineers on the Chickahominy." In *The Richmond Campaign of 1862: The Peninsula and the Seven Days*, edited by Gary W. Gallagher, 44–65. Chapel Hill: University of North Carolina Press, 2000.

———, comp. "The Grand Campaign: A Journal of Operations on the Peninsula, March 17–August 26, 1862." In *The Peninsula Campaign of 1862: From Yorktown to the Seven Days*, vol. 1, edited by William J. Miller, 177–206. El Dorado, CA: Savas, 2013.

———. "Weather Still Execrable: Climatological Notes on the Peninsula Campaign, March through August 1862." In *The Peninsula Campaign of 1862: Yorktown to the Seven Days*, vol. 3, edited by William J. Miller, 184–92. Campbell, CA: Savas, 1997.

Mink, Eric. "Can't See the Battlefield for the Trees: Lost Viewshed from Lee's Command Post" (February 25, 2011). https://npsfrsp.wordpress.com/2011/02/25/cant-see-the-battlefield-for-the-trees-lost-viewshed-from-lees-command-post/.

Monette, Roland, and Stewart Ware. "Early Forest Succession in the Virginia Coastal Plain." *Bulletin of the Torrey Botanical Club* 110, no. 1 (January–March 1983): 80–86.

Montague, Ludwell Lee. "Subsistence of the Army of the Valley." *Military Affairs* 12 (October 1948): 226–31.

Morgan, Oliver. "Infectious Disease Risks from Dead Bodies following Natural Disasters." *Rev Panam Salud Publica / Pan Am Journal of Public Health* 15, no. 5 (2004): 307–12. http://publications.paho.org/pdf/dead_bodies.pdf.

Nash, Linda. "The Agency of Nature or the Nature of Agency?" *Environmental History* 10 (January 2005): 67–69.

National Park Service. "Cultural Landscapes Inventory, 2017: Malvern Hill, Richmond National Battlefield Park." http://www.npshistory.com/publications/rich/cli-malvern-hill.pdf. Accessed September 9, 2018.

———. "Frequently Asked Questions." https://www.nps.gov/gett/faqs.htm. February 9, 2018.

"Natural Climate Oscillations." http://climap.net/natural-climate-oscillations. Accessed October 7, 2014.

"Nature & Science." Andersonville National Historic Site. https://www.nps.gov/ande/learn/nature/index.htm. Accessed January 25, 2017.

Neil, Bill. "'Friends' Take Active Role in Park Support." *Gettysburg (PA) Times*, July 3, 1993.

Nelson, Megan Kate. "'The Difficulties and Seductions of the Desert': Landscapes of War in 1861 New Mexico." In *The Blue, the Gray, and the Green: Toward an Environmental*

History of the Civil War, edited by Brian Allen Drake, 34–51. Athens: University of Georgia Press, 2015.

NOAA (National Oceanic and Atmospheric Administration). "Earth System Research Laboratory, Physical Sciences Division: Atmospheric River Questions and Answers." http://www.esrl.noaa.gov/psd/atmrivers/questions/. Accessed October 6, 2014.

———. "What Is the Pineapple Express?" https://oceanservice.noaa.gov/facts/pineapple-express.html. Accessed May 14, 2019.

Noe, Kenneth W. "The Drought That Changed the War." *New York Times*, Opinionator blog, October 12, 2012. http://opinionator.blogs.nytimes.com/2012/10/12/the-drought-that-changed-the-war/.

———. "Heat of Battle: Climate, Weather, and the First Battle of Manassas." *Civil War Monitor* 5 (Fall 2015): 54–63, 76.

"North Alton Confederate Cemetery, Alton, Illinois." https://www.nps.gov/nr/travel/national_cemeteries/illinois/North_Alton_Confederate_Cemetery.html. Accessed August 2, 2018.

Occhiuti, Andrea, and Taylor McLelland. "Typhus Overview." *Colonial Disease Digital Textbook*. https://colonialdiseasedigitaltextbook.wikispaces.com/5.+Typhus. Accessed November 2, 2015.

Onion, Rebecca. "Late Nineteenth-Century Maps Show Measles Mortality before Vaccines." *Slate*, February 3, 2015. http://www.slate.com/blogs/the_vault/2015/02/03/history_of_measles_mortality_maps_from_a_time_before_vaccines.html.

Pfanz, Don. "Burying the Dead at Spotsylvania." Mysteries & Conundrums, April 4, 2011. https://npsfrsp.wordpress.com/2011/04/04/burying-the-dead-at-spotsylvania-1864/.

Phillips, Gervase. "Writing Horses into American Civil War History." *War in History* 20, no. 2 (April 2013): 160–81.

Porter, Keith, et al. "Overview of the ARkStorm Scenario." Reston, VA: U.S. Geological Survey, 2011, 1–2. http://pubs.usgs.gov/of/2010/1312/of2010-1312_text.pdf.

Potter, David M. "Jefferson Davis and the Political Factors in Confederate Defeat." In *Why the North Won the Civil War*, edited by David Herbert Donald, 93–114. Baton Rouge: Louisiana State University Press, 1960. Reprint, New York: Simon & Schuster, 1996.

Powell, Harrison. "'Seven Year Locusts': The Deforestation of Spotsylvania County during the American Civil War." *Essays in History* 44 (2010). http://www.essaysinhistory.com/seven-year-locusts-the-deforestation-of-spotsylvania-county-during-the-american-civil-war/.

Pyle, G. F. "The Diffusion of Cholera in the United States in the Nineteenth Century." *Geographical Analysis* 1 (January 1969): 59–75.

Ramsdell, Charles W. "General Robert E. Lee's Horse Supply, 1862–1865." *American Historical Review* 3, no. 4 (July 1930): 758–77.

Reardon, Carol. "A Hard Road to Travel: The Impact of Continuous Operations on the Army of the Potomac and the Army of Northern Virginia in May 1864." In *The Spotsylvania Campaign*, edited by Gary W. Gallagher, 170–202. Chapel Hill: University of North Carolina Press. 1998.

Reyes, David. "Dry Years of 1862–65 Changed O.C. Life." *Los Angeles Times*, May 6, 1991.

Roberts, Jacob. "Yellow Fever Fiend." *Distillations*, Science History Institute (Spring 2014). https://www.sciencehistory.org/distillations/magazine/yellow-fever-fiend.

Rubin, Julius. "The Limits of Agricultural Progress in the Nineteenth-Century South." *Agricultural History* 49, no. 2 (April 1975): 362–73.

Sachs, Aaron. "Stumps in the Wilderness." In *The Blue, the Gray, and the Green: Toward an Environmental History of the Civil War*, edited by Brian Allen Drake, 96–112. Athens: University of Georgia Press, 2015.

Schmidt, James M. "Six Feet Under." *Civil War News*, July 2011. http://civilwarmed.blogspot.com/2011/06/medical-department-39-body-bags-and.html.

Sharrer, G. Terry. "The Great Glanders Epizootic, 1861–1866." *Agricultural History* 61, no. 9 (Winter 1995): 79–98.

Shulten, Susan. "Mismapping the Peninsula." *New York Times*, Disunion blog, April 20, 2012. http://opinionator.blogs.nytimes.com/2012/04/20/mismapping-the-peninsula/?_r=0.

Silver, Nate. "The Weather Man Is Not a Moron." *New York Times Magazine*, September 7, 2012. http://www.nytimes.com/2012/09/09/magazine/the-weatherman-is-not-a-moron.html?pagewanted=all.

Skocpol, Theda. "America's First Social Security System: The Expansion of Benefits for Civil War Veterans." *Political Science Quarterly* 108, no. 1 (March 1993): 85–86.

Smallman-Raynor, Matthew, and Andrew D. Clift. "The Geographical Spread of Cholera in the Crimean War: Epidemic Transmission in the Camp Systems of the British Army of the East, 1854–1855." *Journal of Historical Geography* 30 (January 2004): 32–69.

Soniak, Matt. "Why Some Civil War Soldiers Glowed in the Dark." http://mentalfloss.com/article/30380/why-some-civil-war-soldiers-glowed-dark. Accessed January 31, 2017.

Southern Memorial Association. "The Civil War Quartermaster's Glanders Stable." http://www.gravegarden.org/civil-war-glanders-stable. Accessed April 10, 2015.

Stewart, Mart. "Walking, Running, and Marching into an Environmental History of the Civil War." In *The Blue, the Gray, and the Green: Toward an Environmental History of the Civil War*, edited by Brian Allen Drake, 209–24. Athens: University of Georgia Press, 2015.

———. "Whether Wast, Deodand, or Stray: Cattle, Culture, and the Environment in Early Georgia." *Agricultural History* 65, no. 3 (Summer 1991): 1–28.

Stroud, Ellen. "Does Nature Always Matter? Following Dirt through History." *History and Theory Theme Issue* 42, no. 4 (December 2003): 75–81.

Sutter, Paul S. "Epilogue: Waving the Muddy Shirt." In *The Blue, the Gray, and the Green: Toward an Environmental History of the Civil War*, edited by Brian Allen Drake, 225–36. Athens: University of Georgia Press, 2015.

———. Foreword to Lisa M. Brady, *War upon the Land: Military Strategy and the Transformation of Southern Landscapes during the American Civil War*, xiii–xiv. Athens: University of Georgia Press, 2012.

———. "The World with Us: The State of American Environmental History." *Journal of American History* 100 (June 2013): 94–119.

Swanson, Drew A. "War Is Hell, So Have a Chew: The Persistence of Agroenvironmental Ideas in the Civil War Piedmont." In *The Blue, the Gray, and the Green: Toward an Environmental History of the Civil War*, edited by Brian Allen Drake, 163–90. Athens: University of Georgia Press, 2015.

Tao, Jing, and Karen Manci. "Estimating Manure Production, Storage Size, and Land Application Area." Agricultural and Natural Resources, the Ohio State University Extension (2008). http://ohioline.osu.edu/aex-fact/pdf/0715.pdf.

Tarr, Joel. *The Search for the Ultimate Sink: Urban Pollution in Historical Perspective*. Akron, OH: University of Akron Press, 1996.

"Tropical Plant Database: Quinine." http://www.rain-tree.com/quinine.htm#.ViPq-3pVhHx. Accessed October 2, 2015.

Villacorta, Natalie. "The Civil War Wounded in Photographs." *Yale Medical Magazine* 47, no. 2 (Winter 2017). http://ymm.yale.edu/winter2013/features/capsule/145467.

"Visible Proofs: Forensic Views of the Body." National Institute of Health, National Library of Medicine. http://www.nlm.nih.gov/visibleproofs/galleries/technologies/blowfly.html. Accessed April 11, 2015.

Wegner, Ansley Herring. "Phantom Pain: Civil War Amputation and North Carolina's Maimed Veterans." *North Carolina Historical Review* 75, no. 3 (July 1998): 277–96.

Weier, John. "El Niño's Extended Family." November 1999. http://earthobservatory.nasa.gov/Features/Oscillations/.

Weiss, Holger. "Dirty Water, People on the Move: Cholera Asiatica and the Shrinking of Time and Space during the Nineteenth Century." In *When Disease Makes History: Epidemics and Great Historical Turning Points*, edited by Pekka Hamalainen, 187–225. Helsinki: Helsinki University Press, 2006.

"What Is the Rain Shadow Effect?" *World Atlas*. https://www.worldatlas.com/articles/what-is-the-rain-shadow-effect.html. Accessed May 14, 2019.

Wilford, John Noble. "How Epidemics Helped Shape the Modern Metropolis." *New York Times*, April 15, 2008.

Williams, David. "Bitterly Divided: Georgia's Inner Civil War." In *Breaking the Heartland: The Civil War in Georgia*, edited by John D. Fowler and David B. Parker, 19–45. Macon, GA: Mercer University Press, 2011.

Wilson, Charles Reagan. "Pork." In *The New Encyclopedia of Southern Culture*, vol. 7. Edited by John T. Edge, 88–91. Chapel Hill: University of North Carolina Press, 2007.

World Health Organization. "Typhus Fever (Epidemic Louse-Borne Typhus)." http://www.who.int/ith/diseases/typhusfever/en/. Accessed November 2, 2015.

Worrall, Simon. "When, How Did the First Americans Arrive? It's Complicated." *National Geographic*, June 9, 2018. https://news.nationalgeographic.com/2018/06/when-and-how-did-the-first-americans-arrive--its-complicated-/.

Worster, Donald. "Transformations of the Earth: Toward an Agroecological Perspective in History." *Journal of American History* 76, no. 4 (March 1990): 1087–106.

Zapata, Mariana. "How Civil War Soldiers Gave Themselves Syphilis." *Slate*, January 4, 2017. https://slate.com/human-interest/2017/01/during-the-civil-war-smallpox-vaccinations-were-hard-to-come-by-so-many-soldiers-took-a-diy-approach.html.

Zax, David. "Civil War Geology." *Smithsonian Magazine*, April 3, 2009. https://www.smithsonianmag.com/history/civil-war-geology-123489220/.

Theses and Dissertations

Brennan, Matthew Philip. "The Civil War Diet." MA thesis, Virginia Polytechnic Institute and State University, 2005.

Burns, Michael. "War and Nature in Northern Virginia: An Environmental History of the Second Manassas Campaign." PhD thesis, Texas Christian University, 2018.

Gerleman, David J. "Unchronicled Heroes: A Study of the Union Cavalry Horses in the Eastern Theater: Care, Treatment, and Use, 1861–1865." PhD diss., Southern Illinois University, 1999.

Mauldin, Erin Stewart. "Unredeemed Land: The U.S. Civil War, Changing Land Use

Practices, and the Environmental Limitations of Agriculture in the South, 1840–1880." Georgetown University, PhD diss., 2014.

Padilla, Jalynn Olsen. "Army of 'Cripples': Northern Civil War Amputees, Disability, and Manhood in Victorian America." PhD diss., University of Delaware, 2007.

Privette, Lindsay Rae. "'Fighting Johnnies, Fevers, and Mosquitoes': A Medical History of the Vicksburg Campaign." PhD diss., University of Alabama, 2018.

Schieffler, George David. "Civil War in the Delta: Environment, Race, and the 1863 Helena Campaign." PhD diss., University of Arkansas, 2017.

INDEX

Adams, Charles Francis, 114
Aedes aegypti, 31. *See also* mosquitoes
African American troops, 140, 143, 160, 188
Alabama, 46, 70, 82; cattle herding in, 127; cultivated land in, 194; loss of livestock in, 191, 195; ore mining in, 111; salt refining in, 162, 163; Union raids in, 66–67, 87, 124
Alabama units: Tenth Regiment Infantry, 23
Albuquerque, NM, 43, 44
Alcott, Louisa May, 154
Alexander, E. Porter, 93
alfalfa, 128
Alton, IL, 36, 37
American Veterinary Medicine Association, 193
Amherst County, VA, 180
amputations, 153, 154
Anaplasma marginale, 128–29
Anaplasmosis, 128
Andersonville prison, 156–57
"Angel's Glow," 149–50, 153
Anopheles quadrimaculatus, 27–29. *See also* mosquitoes
Antietam, Battle of, 6, 77–79, 112, 166; battlefield preservation of, 184–85; burning horses after, 119
Apache Indians, 39
Appomattox, VA, 7, 187, 190, 193, 196
Arizona, 15, 39, 43, 44
Arkansas, 14, 88; cut off from rest of Confederacy, 101; hog cholera in, 126; loss of livestock in, 191–92; salt refining in, 163
ARkStorm, 42–43
Army of Northern Virginia, 72, 155; at Battle of the Wilderness, 141; at Fredericksburg, 183; on Gettysburg campaign, 93, 102, 115; measles in, 11; shortage of food for, 92
Army of Tennessee (Confederate), 106, 115
Army of Tennessee (Union), 140
Army of the Cumberland, 115, 166
Army of the Mississippi (Confederate), 67
Army of the Ohio (Union), 66–67
Army of the Potomac: 12, 29, 140, 152; in Antietam campaign, 76; at Battle of Chancellorsville, 92; at Battle of Gettysburg, 167; at Battle of Spotsylvania, 133; at Battle of the Wilderness, 141, 170; health of, 11, 35, 64; horses in, 109, 112, 114; in "Mud March," 1–5; in Peninsula Campaign, 52, 57, 74
Army of the Tennessee (Union), 49, 96
Army of Virginia (Union), 72, 74
Atlanta, GA, 7; destruction of, 175; food riots in, 86; founding of, 172–73; horses impressed from, 116; Sherman attacks, 140–41, 173–74; water supply in, 173–74
atmospheric river, 42

Babesia, 128
Babesiosis, 128–29
bacon, 13, 86, 92, 121, 130; calories in, 206n13; destroyed, 88, 92, 124; imported from

England, 125; price of, 87; shortage at Vicksburg, 99–100; traded for cotton, 125
Banks, Nathaniel, 73, 140
Barksdale, William, 23
barley, 128
Barnard, John G., 53
Baton Rouge, LA, 81, 94
Baylor, John, 15, 43
Beaufort, NC, 32
Beauregard, P. G. T., 36, 67
beef, 13, 84, 99, 121, 191; as food for armies, 126, 129–31; calories in, 206n13; northern exports of, 89–90
Bell, Andrew McIlwaine, 30, 32
Berry, Stephen, 6
Billings, John Shaw, 189
Birmingham, AL, 111
Blackburn, Luke Pryor, 33–34
blockade: curtails imports of leather, 131; curtails imports of metals, 111; curtails imports of quinine, 30; curtails imports of salt, 84, 163
blowflies, 117–18, 150, 186
Booth, John Wilkes, 34
Borden, Gail, 90
Boston, MA, 23
Brady, Lisa M., 4, 124, 174
Bragg, Braxton, 67–70, 115–16
Brandy Station, 153; Battle of, 103–5, 111
Brewer, William, 40
Bridgeport, AL, 116, 117
Bruinsburg, MS, 97
Buckner, Simon Bolivar, 48
Buell, Don Carlos, 49, 66–68, 70, 115
Bull Run: First Battle of, 12, 14, 150; Second Battle of, 74, 76, 152
Burbridge, Stephen G., 160, 161, 165
Burden, Henry, 111
Burnside, Ambrose: at Antietam, 77, 166, 184; in "Mud March," 1–3, 5, 92
Burr, Richard, 148
Butler, Benjamin F., 32, 33, 140

Cairo, IL, 11
Calhoun, James A., 173
California, 6, 24, 38; Confederate designs on, 15, 43; drought in, 66, 131, 191; flooding in, 39–40, 42, 46, 50; national parks in, 197; price of cattle in, 209n3
"California Column," 41–43
Callaway, Joshua, 49, 50
calomel, 21–22
Camp Moore, 11
Cashtown, PA, 102
casualties: in Atlanta campaign, 141; at Brandy Station, 104; at Chancellorsville, 93; at Chickamauga, 116; at Cold Harbor, 142; at the Crater, 143; in Grierson's raid, 98; at Malvern Hill, 64; in Overland Campaign, 143–44; at Petersburg, 142; in prisoner of war camps, 155–56; at Saltville, 161; at Second Battle of Bull Run, 74; in Seven Days' battles, 61; at Seven Pines, 60; at Shiloh, 49; at Spotsylvania, 134, 142; at the Wilderness, 141
cattle, 5, 6, 76, 97, 164; in Army of the Potomac, 62; butchered after Antietam, 72, 79; butchered after Gettysburg, 102; deaths in California, 40, 131, 191; deaths in Texas, 131; diet of, 126–28; diseases of, 127–28, 193; drowned along Mississippi River, 81; fenced in pastures, 195; as food supply for armies, 103, 123, 124; killed at Vicksburg, 98, 100; leather from, 129, 131; losses during war, 191–92; in North, 88–89; numbers of, 127, 129; price of, 209n3; recaptured by Union forces, 219n69; reproduction of, 121; seized by armies, 130; as winter food, 83–84
Cedar Mountain, Battle of, 119
Cedar Run, Battle of, 72
Centers for Disease Control, 35
Chancellorsville, Battle of, 92–93
chaos theory, 51
Charleston, SC, 32
Charleston, WV, 162
Chattanooga, TN, 46, 66, 67, 123; Confederate prisoners at, 151; siege of, 116–19
Chicago, IL, 78, 89, 122, 167, 213n22
Chickahominy River, 53, 59–61, 63–65, 182
Chickamauga, Battle of, 116, 178

Chimborazo hospital, 155
cholera, 24–25, 31, 37, 80; epidemic in New York, 207n39; epidemic in Buffalo, NY, 213n22; epidemic in Chicago, 213n22
Cincinnati, OH, 67, 107, 122, 162
Clausewitz, Carl von, 69, 70
Clostridium, 153
coffee, 13, 14, 29, 62, 85, 214n37
Cold Harbor, Battle of, 142
Comanche Indians, 44, 46
conciliation, 71–72
condensed milk, 90, 91
Confederate Commissary Department, 91, 125
Confederate Congress, 83, 86, 87
Connecticut units: Third Regiment Infantry, 14; Sixteenth Regiment Infantry, 156
Conness, John, 197, 199
Conscription Act, 83–84
Corinth, MS, 49–50, 66, 67, 96
corn, 14, 53, 71, 75–77; decreased production of, 94, 180–81, 194; as dietary staple, 81, 168; exports of, 90; as food for hogs, 121, 122, 192; as food for horses, 110, 115; impressed by armies, 92, 97, 124; poor crop of, 82, 83, 85; seized in food riots, 86–88; shortage at Vicksburg, 98
cotton, 82, 83, 94, 180, 181; causes soil erosion, 196; exports of, 195; planted instead of food crops, 85; traded for food, 125; traded for horses, 107
cowpox, 35–36
Cram, Thomas Jefferson, 54
Crater, Battle of the, 142–43
Crimean War, 25
Crosby, Alfred W., 4
Culpeper County, VA, 71–75, 77
Cumberland River, 46–48

Danville, VA, 151, 180, 196
Davidson, NC, 111
Davis, Jefferson, 75, 93, 94, 115; deals with food riots, 86; refuses to trade cotton for food, 125
dead bodies: burial of, 137–38, 145–46; decomposition of, 136, 138, 139, 143; embalming of, 146–47; odor of, 145–46; preservation of, 144–45; transport of, 145
deaths: from amputations, 153–54; burial problems, 50, 80; from dehydration, 16; from diarrhea and dysentery, 19; effects on labor supply, 194; effects on marriage rates, 193–94; insects appear after, 138; from measles, 11; in prisons, 155; at Spotsylvania, 134; total number of, 136, 193; from typhus, 26; from wounds, 153
Deere, John, 89
DeGruccio, Michael, 182
dehydration, 13–16, 24, 37, 62
desertion, 5, 125
Devine, Shauna, 189
diarrhea, 14, 18–19, 29, 37, 186; in Antietam campaign, 76, 80; attempted remedies for, 21–22; caused by contaminated water, 155; deaths from, 18; in Peninsula Campaign, 62–65; as symptom of cholera, 24; as symptom of typhoid fever, 16; among veterans after the war, 189
diphtheria, 11–12
disabled veterans, 155, 158–59
Dodge, Grenville, 87, 124
Dodge, Jacob Richard, 193
Dodge, Theodore A., 2, 3
Downs, Jim, 188
dysentery, 18–21, 29, 37, 118; at Andersonville prison, 156; in Peninsula Campaign, 62–65; after Shiloh, 50

E. coli, 18, 19
Edgecombe County, NC, 9
Ehlen, Judy, 168
Ellsworth, Elmer, 147
Elmira prison, 155
El Niño, 41–42
El Niño Southern Oscillation (ENSO), 41–42, 50
Emerson, Ralph Waldo, 199
Emory and Henry College, 161
Ewell, Richard, 133

Fair Oaks, Battle of, 119, 120, 138
farcy, 112–13, 120
farming: mechanical implements for, 89;

practices in South, 71, 82–83, 122, 196; production in North, 88–89; use of fertilizer in, 83
Faust, Drew Gilpin, 136, 138, 147
Ferguson, Champ, 161
Fiege, Mark, 215n56
Fisk, Wilbur, 135
Florence, AL, 48
Florida, 130, 192
Floyd, John, 48
food: canning of, 90; Confederate army captures Union supplies of, 74; Confederate army collects, 86–87, 92; decreased production of, 194; in Maryland, 75; prices of, 87, 98–99; production of rations, 91; riots over, 86; shortages in prisons, 156; spurs Confederate invasions, 75, 93; Union forces confiscate, 72, 73, 79; Union forces destroy, 87–88, 124. *See also* bacon; beef; coffee; corn; pork; rice; wheat
forests: destruction of, 178, 179, 195; growth in cleared lands, 53, 169; losses compared to other wars, 195; as postwar inspiration, 197; stripped for military use, 175–76; used for open range grazing, 127; used to fuel salt furnaces, 164–65
Forrest, Nathan Bedford, 48, 140
Fort Donelson, 47–48, 50, 70
Fort Fillmore, 12, 15, 43
Fort Henry, 47, 48, 50, 70
Fort Monroe, 52, 54, 55
Fort Pillow, 36, 37; massacre at, 140, 143
Fort Sumter, 105, 123
Fort Yuma, 39, 41
Fredericksburg, VA, 1, 2, 52, 92; Battle of, 5; battlefield preservation in, 183; deforestation near, 177
Freemon, Frank, 208n44

Gaines Mill, Battle of, 61, 182
gangrene, 21, 153
Garysburg, NC, 9, 10, 12, 16
Georgia, 85, 115–16, 130; crop reductions in, 194; food riots in, 86; livestock losses in, 191–92, 195; salt refining in, 163; Sherman's campaign in, 140, 173; soil erosion in, 181
Georgia units: Sixth Regiment Infantry, 18
Gettysburg, PA, 94, 139, 145; Battle of, 102–4, 167–68, 178; battlefield preservation of, 185; burials at, 139; Confederate march to, 131; Confederate retreat from, 114–15, 130; dealing with dead horses at, 102, 117, 119–20
Gettysburg National Cemetery, 134
Giesboro Point, 107–8, 120
glanders, 112–13, 120, 125–26, 193. *See also* horses, diseases of
Glendale, Battle of, 61
Glorieta Pass, Battle of, 43–44
gonorrhea, 22–24, 188. *See also* venereal disease
Gordon, George L., 16
Gordon, John B., 12
Grand Junction, TN, 96
Grant, Ulysses S., 11, 70, 124, 139, 199; at Appomattox, 187, 190, 191; captures Fort Donelson, 48–49; captures Fort Henry, 47; named general-in-chief, 140; in Overland Campaign, 141–43, 153; at Shiloh, 49–50; signs weather service legislation, 190; takes over at Chattanooga, 117; at the Wilderness, 141, 170–71; in Vicksburg campaign, 95–98
greased heel, 112. *See also* horses, diseases of
Great Appalachian Valley, 165–66
Great California Flood, 46
Greenville, AL, 84
Greenville, LA, 108
Grierson, Benjamin, 97, 214n37
guano, 83
guerrillas, 88

Hagerstown, MD, 76, 130
Halleck, Henry, 11, 49, 50, 66, 139
Hammond, William A., 152
Harpers Ferry, VA, 76
Harris, David Golightly, 81, 82, 83
Harris, Nathaniel H., 133, 134, 219n2
Harrisburg, PA, 75, 108, 167

Harrison's Landing, VA, 61, 63–66
heat exhaustion, 13, 14, 15
heat stroke, 13
Helena, AR, 88
Henrico County, VA, 179–80
Hill, D. H., 83
Hilton Head, SC, 32
hog cholera, 125–26, 192, 193
hogs, 5, 6, 72, 97; after Antietam, 79; diet of, 121; diseases of, 125–26, 192, 193; drives of, 122; eat dead bodies, 137, 139; imported in South, 123, 125; losses of, 192, 224n15; in North, 89; reproduction of, 121; as scavengers, 103; shortage of, in South, 123–25; size of, 121, 122; traded for cotton, 125; as winter food, 83–84
Holly Springs, MS, 96
Holmes, Thomas, 145, 146, 147
Holston River, 46, 160
Holt, Daniel, 139
Homestead Act of 1862, 195
Hood, John Bell, 77, 93, 141
Hooker, Joseph, 2, 5, 92, 93
horses: abandoned, 115; after Antietam, 79; burning of dead, 80, 119–20; carcasses of, 5, 117; Confederate depots for, 111; Confederate foraging for, 110; Confederate impressment of, 107, 114, 116; contracted by U.S. government, 107–9; desperate for water, 44; dying of exhaustion, 62; dead at Brandy Station, 104; dead at Chattanooga, 117; dead at Gettysburg, 102–3; development of, 105; diet of, 109–10; diseases of, 112, 113, 217n25; disposal of dead, 118–19; as engines, 109; in Giesboro Depot, 107; as health hazard, 117–19; horseshoes for, 111, 115; inadequate supply of, 106; killed at Chickamauga, 116; killed by U.S. forces, 107; lack of food for, 41, 44, 114–15; life expectancy in Confederate army, 114; losses of, 191; numbers of, in the country, 105; poison water sources, 119; prices for, 107; rehabilitation of, 108; struggling in mud, 2, 5, 6; Union impressment of, 114; used as fertilizer, 120; value in death, 120; veterinary care for, 113; waste from, 79–80
hospitals, 23, 24, 30, 149; at Andersonville, 156; at Brandy Station, 153; derided by soldiers, 154; development and improvement of, 155, 189; at Fredericksburg, 153; maggots in, 150, 151, 156; murder of soldiers in, 161; number of, 155; in Peninsula Campaign, 63; rats in, 152; at Richmond, 155; at Sharpsburg, 79, 119
Hottel, Zachary, 211n46
Humphreys, Margaret, 18, 26
Hunter, Alexander, 76
Hussey, Obed, 89

Illinois units: Eighth Regiment Cavalry, 138
impressment law, 86, 87
influenza, 31
Iowa State University, 193
Iowa units: First Regiment Infantry, 15
Iuka, MS, 96

Jackson, Andrew, Jr., 36
Jackson, MS, 96, 97, 98, 124
Jackson, Stonewall, 72, 110, 172; in Antietam campaign, 76; at Chancellorsville, 92–93; at Second Battle of Bull Run, 74; in Shenandoah Valley campaign, 61, 73, 166
James River, 7, 52, 53, 59
Jamestown, VA, 17, 18
jaundice, 31
Jenner, Edward, 35
Johnston, Albert Sidney, 47, 49
Johnston, Joseph E., 65, 98, 113; and Atlanta campaign, 140; corresponds with Pemberton, 216n69; in Peninsula Campaign, 56, 59–60; pulls back from Manassas, 52
Jones, John B., 86, 87
Jones, Jonathan S., 189
Jones, Joseph, 151
Jornada del Muerto, 15

Kearny, Phil, 65
Kellogg, Robert, 156

Index

Kentucky, 46, 49, 50, 160, 164; Confederate invasion of, 66–69; hogs in, 123; horse breeders in, 195; production of tobacco in, 180; salt refining in, 162–63; under Union control, 85, 123–25, 140
Kentwood, LA, 11
Keyes, Erasmus D., 54–56, 64
King, Thomas Starr, 197
Kirby, Jack Temple, 3, 122, 196
Knox, Thomas, 15
Knoxville, TN, 46, 122, 123
Krick, Robert K., 51, 219n2

Lagrange, TN, 97
land: decrease in cultivation of, 159, 194; effects of emancipation on labor for, 181; erosion of, 181, 196; loss of cultivated acres of, 85, 101; maintenance of, 181, 196; soil exhaustion of, 196
La Niña, 41, 42, 46, 50, 51, 66
Lapham, Increase A., 190
leather, 129–31, 162
Ledford, Brian, 222n11
Lee, Robert E., 1, 20, 34, 153, 199; in Antietam campaign, 75–79; at Appomattox, 187, 190–91; concern for horses, 105, 107, 110, 114–15; forced to go on defensive, 115; at Fredericksburg, 183; in Gettysburg campaign, 93–94, 102–3, 114, 130, 167–68; and his horse, 109; impresses leather, 131; laments lack of food for army, 91–92; in Overland Campaign, 140–44; in Peninsula Campaign, 56, 60–65, 70, 166; requests Longstreet's return before Chancellorsville, 215n54; in second Bull Run campaign, 72, 74; at Spotsylvania, 133; at the Wilderness, 170–72
Leister, Lydia, 102–4, 117, 119–20
Letterman, Jonathan, 63–64, 152
Lexington, KY, 67, 164
lice, 25, 26, 208n44
Lincoln, Abraham, 12, 34, 94, 142; approves Grant's plans, 140; delivers Gettysburg Address, 139; ends Peninsula Campaign, 64–66; interactions with Burnside, 1, 2, 92; issues preliminary Emancipation Proclamation, 78; praises Rosecrans, 167; relationship with McClellan, 52–53, 61, 74, 112; relieves Rosecrans of command, 117; shifts from conciliation, 71–72; signs bill establishing Yosemite, 199; stops cotton trade with South, 125
Little Ice Age, 50, 51
Liutkus-Pierce, Cynthia, 211n49
Longstreet, James, 74, 115; at Chickamauga, 116; collects food for Lee's army, 91–92, 110; at the Wilderness, 170–72
Lorenz, Edward, 51, 69
Los Angeles, CA, 39, 191
Loughborough, Mary, 98, 99
Louisiana, 81, 85, 101, 140; crop losses in, 194; livestock numbers in, 192; salt refining in, 162
Louisville, KY, 23, 67, 68, 91
Lowry, Thomas, 188
Luray, VA, 72
Lynchburg, VA, 111, 113, 180
Lynde, Isaac, 15
Lyon, Nathaniel, 12, 14–15

Maccrady formation, 162, 163, 222n7
Mackee, C. B., 51
Madden-Julian Oscillation (MJO), 42
maggot debridement therapy, 152
maggots, 117, 138, 150–53, 156, 186
Magruder, John, 16, 18, 54–56
malaria, 27–32, 37, 128; in Peninsula Campaign, 63, 66; in Vicksburg campaign, 99
Malvern Hill, 7, 182; Battle of, 61, 64–66
Manassas, VA, 12, 14, 52, 74, 110, 113
March to the Sea, 174
Marengo County, AL, 181
Mariposa Grove, 197, 199
Marsh, George Perkins, 199
Maryland, 37, 112, 180, 184; Confederate invasion of, 75–79, 90; horse breeders in, 195
Massachusetts, 189
Massachusetts units: Ninth Regiment Artillery, 104
Mather, Stephen, 199
Mauldin, Erin Stewart, 126, 179, 181

McClellan, George B., 29, 104, 113, 118, 190; and his horse, 109; in Antietam campaign, 76–77, 112; in Peninsula Campaign, 52–66, 70–72, 81, 90, 105, 165–66
McClellan, Henry, 104
McCollough, Ben, 14
McCormick, Cyrus, 89
McCurry, Stephanie, 86
McDowell, Irvin, 12, 14
McGee, W. J. 16
McLaws, Lafayette, 55
McLean, Wilmer, 187
McParlin, Thomas A., 153
McSween, Murdoch, 177
Meade, George Gordon, 102, 140
measles, 10–12, 17, 23, 34, 37, 188
medical care: ambulance corps, 152; antebellum doctors, 147; field hospitals, 152; improvements in, 189; surgeons in war, 152; tending to wounded, 149
medicines, 21–23
Meier, Kathryn Shively, 20
Meigs, Montgomery C., 107, 108, 109, 114, 130
Melville, Herman, 7
Memphis, TN, 88, 96, 125, 140
mercurial gangrene, 21
mercury, 21
metabolic equivalent expenditures, 211n46
Mexico, 107, 131
Miller, David R., 77, 166, 184
Milliken's Bend, LA, 18
Milwaukee, WI, 23, 79
Minié, Claude Étienne, 148
Minnesota units: First Regiment Infantry, 14
Mississippi, 49, 82, 130; cattle herding in, 127; crop losses in, 194; disease in, 18; food confiscated in, 124; livestock numbers in, 192; Union forces in, 66, 91, 93, 96–97
Mississippi River, 36, 37, 72, 107, 214n35; flooding of, 81; guerrilla raids along, 88; Union forces control, 124; in Vicksburg campaign, 94, 99
Missouri State Guard, 12

Mitchell County, NC, 82
Mobile, AL, 67, 86
Mojave Desert, 39
Montgomery, AL, 131
Morris, Francis, 106
Morris, Roy, Jr., 21
mosquitoes, 27–34, 37, 63, 64
mud, 49, 50, 69; in Chattanooga siege, 117; in Mud March, 2–3, 5, 92; in Peninsula Campaign, 54, 56–62
Mud March, 1–3, 5, 205n10
Muir, John, 198
mules, 2, 44, 79, 195; in Chattanooga siege, 117; diet of, 109; losses of, 191; numbers of, 105, 109; in Peninsula Campaign, 58–59, 62; in Vicksburg campaign, 100. *See also* horses
mumps, 11
Museum of the Middle Appalachians, 221n3
Myer, Albert J., 190

Nashville, TN, 46, 48, 49, 116; horse depot at, 108; meatpacking plant in, 123; prostitutes in, 23–24
Nassau, Bahamas, 32, 125
Natchez, MS, 94
National Park Service, 182–84, 199
National Weather Service, 190
Nelson, Megan Kate, 16, 179, 195, 210n44
New Albany, MS, 97
New Bern, NC, 32, 66, 88
New Mexico, 12, 15, 39, 41–44, 52, 70
New Orleans, LA, 24, 32, 94, 214n35
New York, NY, 24, 51, 107, 145
New York units: Sixth Regiment Artillery, 178; Second Regiment Infantry, 14; Seventh Regiment Infantry, 20; Eleventh Regiment Infantry, 14; Thirteenth Regiment Infantry, 14; Fifteenth Regiment Infantry, 60; Thirty-Fifth Regiment Infantry, 20; 110th Regiment Infantry, 2
Norfolk, VA, 91
Norfolk formation, 58
North Carolina, 9, 83, 85, 88; livestock losses in, 191; loss of cultivated land in,

194; production of tobacco in, 180, 181; as provider of horses, 107, 110; salt refining in, 163

North Carolina units: Fifteenth Regiment Infantry, 206n1; Fifth State Volunteers, 9, 16, 17

Northrop, Lucius, 87, 123, 125

oats, 53, 81, 89, 102, 128

Ohio River, 46, 66, 68, 90, 126

Ohio units: Twelfth Regiment Cavalry, 161

Olmsted, Frederick Law, 81, 197–99

opium, 21, 189

Overland Campaign, 6, 141–44, 153, 155, 197

Padilla, Jaylynn, 158

Paducah, KY, 46

Paine, Halbert E., 190

Paré, Ambroise, 150

Pemberton, John C., 96–98, 100, 216n69

Peninsula Campaign, 52–66

Pennsylvania, 75, 76, 89, 103, 139; Confederate invasion of, 93–94, 114–15, 130, 167; Confederate retreat from, 114

Perryville, KY, 115, 119; Battle of, 68–69, 70

Petersburg, VA, 86; Battle of, 142

Philadelphia, PA, 107

Pickett, George, 102, 168, 185

pineapple express, 42

Pitts, Hiram, 89

Pitts, John, 89

Pittsylvania County, VA, 180, 181

Plasmodium falciparum, 27. *See also* malaria

Plasmodium vivax, 27. *See also* malaria

Pleasonton, Alfred, 104

pneumonia, 11, 28, 35, 48, 155

Polk, James K., 24

Pope, John C., 72–74, 110

pork, 124, 126, 162, 192; as dietary staple, 81, 84–85, 121–22, 129; northern exports of, 90, 123; theft of, 125; as U.S. army ration, 13

Porter, David, 97

Porter, Horace, 134

Port Hudson, LA, 124

Potomac River, 52, 74–77, 107

prostitutes, 23–24

Pseudomonas mallei, 113. *See also* glanders

quinine, 29, 30, 31

Raleigh, NC, 11

Randolph, George, 125

Randolph County, AL, 86

Rapidan River, 92

Rappahannock River, 52, 92, 103, 141, 177, 183; as location of Mud March, 1, 2, 5

rice, 13, 84, 85, 92, 99

Richmond, KY, 67

Richmond, VA, 1, 5, 73, 91, 140, 179; battlefields near, 182, 183; bread riot in, 86; disease in, 11, 23; firewood in, 164; food shortages in, 86–88, 125; hogs sent to, 126; hospitals in, 152, 155; in Overland Campaign, 142; in Peninsula Campaign, 52–61, 65–66, 70, 72, 165–66; shoe factory in, 131; tobacco production in, 180; U.S. cavalry raids on, 111

Rickettsia prowazekkii, 25. *See also* typhus

Ringwalt, Samuel, 106, 113

Rio Grande, 43, 44, 125

Rocky Mount, NC, 9

Rosecrans, William, 115–17, 166

Rubeola, 10, 11, 188. *See also* measles

Ruffin, Edmund, 82

rye, 128

Sachs, Aaron, 197

Sacramento, CA, 39, 40

Saint Quentin, Battle of, 151

Salisbury, NC, 86

Salmonella enterica, 17. *See also* typhoid fever

Salmonella paratyphi, 17. *See also* typhoid fever

Salmonella typhimurium, 17. *See also* typhoid fever

salt, 7, 13; for horse diet, 109; as preservative, 84–85, 121; production of, 162–65, 172, 174

saltpeter, 222n14

Saltville, VA, 161–65, 221n3

Saltville Fault, 222n7

San Antonio, TX, 43, 44
San Augustin Spring, NM, 15
San Francisco, CA, 39
Santa Ana River, 39
Santa Fe, NM, 43
Santa Fe Trail, 43
Savage Station, Battle of, 61
Savannah, GA, 164
Scollay, G. W., 145
scurvy, 63, 64, 91; at Andersonville prison, 156
Sears, Stephen, 65
Seddon, James, 110, 124
Seven Days' Battles, 61–64, 118, 182
Seven Pines, Battle of, 59–61
Seven Years' War, 36
Sharpsburg, MD, 77–81, 98, 99, 166; battlefield preservation of, 184
Sharrer, G. Terry, 11, 63, 193
Shenandoah Valley campaign, 73, 166
Sherman, William, 160; in Atlanta campaign, 140–41, 173–75; in South Carolina, 177; in Vicksburg campaign, 94, 96–98, 124, 215n63
Shigella, 18, 19
Shiloh, Battle of, 49–50, 149–50
Sibley, Henry Hopkins, 43, 44, 46, 70
Sierra Nevada, 15, 43
Sigel, Franz, 140
Simpson, Tally, 93
smallpox, 34–37, 188
Smith, Edmund Kirby, 67, 125
Smith, William Sooy, 108
Smyth County, VA, 160
Sneden, Robert Knox, 62
Soil Conservation Service, 196
Solomon, Isaac, 90
Sonora, CA, 39
South Carolina, 29, 32, 85, 147; livestock losses in, 191–92, 195, 224n15; loss of cultivated land in, 194
South Mountain, Battle of, 76, 139
Spotsylvania, Battle of, 133–35, 142, 153; wounded soldiers at, 149
Spotsylvania County, VA, 168–69, 179–80
Springfield, MO, 12, 14, 15

Stanton, Edwin, 57
Staphylococci, 153
starvation, 42, 75; in Culpeper County, 72; at Elmira, 155; on home front, 83, 86, 88; in Lee's army, 91; after Second Battle of Bull Run, 152; at Vicksburg, 99–100
stem rust, 81
Stevenson's Depot, AL, 67
Stiles, Robert, 114
St. Louis, MO, 11, 15, 24, 107, 145
Stone, Kate, 85
Stoneman, George, 165
Stones River, Battle of, 115, 166–67
Streptococci, 153
Stroud, Ellen, 4
Stuart, James Ewell Brown (Jeb), 61, 104
Swanson, Drew A., 180, 196
syphilis, 22–24, 36, 188. *See also* venereal disease

tax-in-kind law, 87
Taylor, William H., 21
Tennessee, 18, 36, 46, 50, 52; Confederate raids in, 124, 125, 140; hogs in, 122–23, 125; horse breeders in, 195; livestock numbers in, 192; Union operations in, 49, 66, 67, 85, 91, 96
Tennessee River, 46–48, 67, 70, 116; dead horses in, 119
Texas, 15, 32, 107; cattle in, 127, 130, 131; Confederate army in, 43, 44; cut off from Confederacy, 101; harsh winter in, 131; horse breeders in, 195; livestock numbers in, 191–92
Thoreau, Henry David, 199
ticks, 127
Tilghman, Lloyd, 47
timothy, 128
tobacco, 53, 71, 82, 169; cultivation of, 180–81, 196; decline in production of, 180; traded for horses, 107
Tredegar Iron Works, 86
Tripler, Charles S., 11, 19–20, 35
Trobriand, Régis de, 2, 63
Troy Iron and Nail Works, 111
Tupelo, MS, 67

Turner, Ted, 185
typhoid fever, 37, 80, 118, 137, 154; Confederate soldiers suffer from, 17–18; different names for, 207n22; in Peninsula Campaign, 62
typhus, 25–26, 37, 208n41

Urbanna, VA, 52
U.S. army rations, 13–14; caloric intake of, 13; dried vegetables in, 14, 63
U.S. Colored Cavalry Regiment, Fifth, 160–61, 164
U.S. Department of Agriculture (USDA), 179, 193
U.S. Sanitary Commission, 79
U.S. Signal Corps, 190

Valverde, Battle of, 43, 44
Vance, Zebulon, 83, 88, 110
Van Dorn, Earl, 96
Variola major, 34. *See also* smallpox
Venable, Charles, 219n2
venereal disease, 22–24, 37, 188, 189
Vibrio cholerae, 24. *See also* cholera
Vicksburg, MS, 81, 85, 94, 130; Battle of, 6, 96–101, 124
Villepigue, John Bordenave, 36
Virginia, 6, 9, 12, 38, 71, 77; burying dead in, 119, 147; Confederate army impresses food in, 91–92, 110; Confederate army retreats to, 114, 115; Confederates launch invasion from, 103, 166; diseases in, 17, 27; food shortages in, 93, 124, 129, 180; forest development in, 169; livestock losses in, 191–92; Mud March in, 1–3; open range grazing in, 122; Overland Campaign in, 141, 143; Peninsula Campaign in, 52–54, 58, 105; production of tobacco, 180, 196; as provider of horses, 107; salt refining in, 163; Union attack on Saltville, 160–61; Union occupation of, 85, 91; wet weather in, 52–54, 81

Wainwright, Charles, 112, 142
Walker, John J., 124
Warren, Gouverneur, 167

Warren County, MS, 94
Warwick River, 54, 55
Washington, D.C., 1, 12, 23, 34, 188; embalming dead bodies in, 147; horse depot at, 107, 108; hospitals in, 155; prostitutes in, 23; Union armies protect, 52, 72; Union forces retreat to, 14, 74; Union soldiers vaccinated in, 35; weather reporting in, 51, 190
Watt, James, 109
weather: "Civil War drought," 42, 50, 66; drought in California, 66, 131; drought in 1862, 81, 82, 96; drought in South, 66, 114; flooding in California, 39–42; flooding in 1862, 96; flooding in Tennessee Valley, 46–47, 49; flooding on Chickahominy River, 59; flooding on Mississippi River, 81; flooding on Ouachita River, 81; flooding on Tennessee River, 70; flooding on Yazoo River, 81; harsh winter in Texas, 131; heat at Wilson's Creek, 14–15; heat in Kentucky, 67–69; heat in New Mexico, 46; heat in Peninsula Campaign, 55; lack of rain at Vicksburg, 99; during "Mud March," 1–3, 5; observations in D.C., 51; rain as salvation to dehydrated troops, 46; rain at Spotsylvania, 134; rain in Kentucky, 69; rain in Peninsula Campaign, 56–61, 65; "rain shadow effect," 43; snow in Albuquerque, 44; snow in Tennessee, 48
West Virginia, 84, 85, 163, 192
wheat, 53, 71, 76, 102, 128, 168; calories in, 206n13; Confederates impress, 91, 92; decreased production of, 81, 179–80, 194; northern exports of, 90; poor crop of, 72, 81, 82, 94; seized in food riots, 86–87, 89
Wheeler, Joseph, 117
Whisonant, Bob, 168
Whitehouse, Alfred D., 150
Whitman, Walt, 19, 156
whooping cough, 11
Wilderness, Battle of the, 141, 170–72, 197; battlefield preservation of, 183; wounded at, 153
Wiley, Bell Irvin, 180

Williamsburg, VA, 56, 58
Williamsport, MD, 131
Wilmington, DE, 108
Wilmington, NC, 26, 32
Wilson, John S., 122
Wilson's Creek, Battle of, 12, 14
Windsor formation, 58
Winslow, Isaac, 90
Wytheville, VA, 111

Yazoo River, 96, 124
yellow fever, 31–34, 37
yellow jack, 32. *See also* yellow fever
Yellowstone National Park, 199
York River, 53, 58, 63
Yorktown, VA, 16, 54, 56, 58, 65
Yosemite Valley, 197–99

Zacharias, J. F., 151